Sustainable Water

ISSUES IN ENVIRONMENTAL SCIENCE AND TECHNOLOGY

EDITORS:

R.E. Hester, University of York, UK
R.M. Harrison, University of Birmingham, UK

TITLES IN THE SERIES:

1: Mining and its Environmental Impact
2: Waste Incineration and the Environment
3: Waste Treatment and Disposal
4: Volatile Organic Compounds in the Atmosphere
5: Agricultural Chemicals and the Environment
6: Chlorinated Organic Micropollutants
7: Contaminated Land and its Reclamation
8: Air Quality Management
9: Risk Assessment and Risk Management
10: Air Pollution and Health
11: Environmental Impact of Power Generation
12: Endocrine Disrupting Chemicals
13: Chemistry in the Marine Environment
14: Causes and Environmental Implications of Increased UV-B Radiation
15: Food Safety and Food Quality
16: Assessment and Reclamation of Contaminated Land

17: Global Environmental Change
18: Environmental and Health Impact of Solid Waste Management Activities
19: Sustainability and Environmental Impact of Renewable Energy Sources
20: Transport and the Environment
21: Sustainability in Agriculture
22: Chemicals in the Environment: Assessing and Managing Risk
23: Alternatives to Animal Testing
24: Nanotechnology
25: Biodiversity Under Threat
26: Environmental Forensics
27: Electronic Waste Management
28: Air Quality in Urban Environments
29: Carbon Capture
30: Ecosystem Services
31: Sustainable Water

How to obtain future titles on publication

A subscription is available for this series. This will bring delivery of each new volume immediately on publication and also provide you with online access to each title via the Internet. For further information visit http://www.rsc.org/Publishing/Books/issues or write to the address below.

For further information please contact:
Sales and Customer Care, Royal Society of Chemistry, Thomas Graham House, Science Park, Milton Road, Cambridge, CB4 0WF, UK
Telephone: +44 (0)1223 432360, Fax: +44 (0)1223 426017, Email: sales@rsc.org

Foreword

In August 2009 the Royal Society of Chemistry (RSC) hosted the 'International Union of Pure and Applied Chemists' 42nd Congress in Glasgow. This congress featured over 50 symposia, which highlighted the positive impact of Chemistry on the world around us.

A key symposium within the congress focused on sustainable water and the impact of climate change, economics, social justice and changes in the population on water resources. This book showcases some of the topics of discussion that featured in the symposium and emphasises the importance of chemical sciences in achieving sustainable water resources on a global scale.

The RSC has considered how chemical sciences can support sustainable water resources in their Water Sustainability Report published in 2007. The report itself focuses on the key role of chemical sciences in supporting the future shape of water management and covers the entire hydrological cycle, paying particular attention to the management of domestic, industrial and agricultural water use, as well as contamination and climate change. The RSC has further developed water resource sustainability into its priority roadmap for chemical sciences. Both documents can be obtained through the RSC web site (www.rsc.org/ScienceAndTechnology/Policy/Documents/water.asp).

At a global level, pressure on water resources is increasing. There are many factors influencing the hydrological cycle and these include increased economic activity of developing and developed nations, urbanisation and increased population, together with human-induced climate change.

Every nation holds the right to advance its economic prosperity but this leads to greater abstraction of water and a higher risk of pollution, all of which has a profound effect on the environment and ecosystem health. Within the UK it is worth noting that Ecosystem Services, which is part of the government's Department of the Environment, Food and Rural Affairs (Defra), has been set up to establish an ecosystem approach that is truly sustainable (see www.ecosystemservices.org.uk). This is aligned to the United Nations "Millennium Ecosystem Assessment" which highlighted the multiple benefits of an ecosystems approach to developing sustainable national policy. Ecosystem Services are developing research projects that will inform decision making at a national and local level within the UK and this all supports and contributes to sustaining water resources. It is clear then, that the key objective of the Ecosystem Services project is to make the link between ecosystem health, market forces and policy decisions.

Relating the ecosystem approach back to water sustainability can be done by considering the main drivers that influence the changes we face and understanding the common links and influences.

In the first instance we should consider environmental changes. This includes climate change, leading to temperature changes, precipitation changes and a build-up of greenhouse gases in the atmosphere. In addition to these factors we also observe that air quality in general is changing. All of these influences have an impact on the frequency of bad weather events, flood risk and drought. As an example, the increase in the temperature of the planet leads to an increase in snow and glacier melt which implies less winter storage of water that leads to a reduction of water availability in spring. Again, this has an impact on the sustainability of micro and macro flora and fauna within the ecosystem.

Climate change also has a profound influence on the ecosystem health, which puts natural resource availability at risk and limits a nation's ability to achieve full growth potential.

Philippe Quevauviller explores the relationship between water sustainability and climate change in more detail in the first chapter. He comments on some of the observations by the Intergovernmental Panel on Climate Change (IPCC) and links this back to the impact and potential effect on the European objectives in moving to sustainable water management where water sources are demonstrably of good quality and of good ecological status. Finally he outlines the key research priorities for Europe in a global context.

In the second chapter, Alan MacDonald and his co-authors highlight the impact of climate change in an African context, exploring some climate change scenarios and their possible outcomes. This chapter then goes on to consider water availability and water stress and makes the link back to better source control and considers changes in demand across the African continent.

Following on from these chapters, Ulrich Borchers and his co-authors discuss the importance of the Water Framework Directive in delivering "good status" for waters in Europe and presents a case study of work carried out to monitor for priority substances in water bodies. This third chapter touches upon some of the analytical challenges that chemists face when developing methods which will allow countries to demonstrate improvements to water quality.

Having discussed environmental change as a driver in the health and sustainability of the ecosystem and water, we must now also consider the changing trend in human behaviour as a driver that influences water sustainability and ecological wellbeing.

There are a number of trends in human behaviour that impact on our environment that should be considered in this debate. These include market forces, population changes, urbanisation, economic growth, price of raw materials, advancements in technology, and the removal and introduction of species within the ecosystem.

Since the industrial revolution, the pace of change on a global scale has increased exponentially, with new technological advances enabling changes to how raw materials are mined, improvement being made in manufacturing processes and increasing the speed at which knowledge can be shared and applied.

This has many effects, one of which is the change in demand within the population that influence markets. In the United Kingdom, as an example, we expect seasonal perishable goods such as fruit and vegetables to be available all year round. The result of this demand is an increase in imports which drives changes in manufacturing and distribution. In the fourth chapter, Jerry Knox and his co-authors explore the implications that these demands have on the water resources of the exporting nations, and the effect that this can have on our water footprint and food security in the longer term. The chapter also touches on the longer term economic impact of the growing reliance on imported food within the UK which can equally be applied to other countries.

It follows that land use and land management play a crucial role within the ecological framework that supports sustainable water. In many countries across the world we are seeing changes to the population and an increase in urbanisation which impact on land use, ecology, and therefore water resource sustainability. We can see an increasing trend in the use of irrigation and desalination in some water stressed areas. New research and technology is influencing farming practices. New methods of working support changes in woodland and forest management. All of these changes influence the ecosystem and have an impact on water resources. It is therefore critical to understand the relationship between all of these factors when assessing new ideas and research topics that will benefit the sustainability approach.

As the priority and availability of water changes, we must also remember that everyone should have the right to a safe source of drinking water. In chapter five, Adrian McDonald and his co-authors look at social justice at three levels within society. Initially they examine social justice at the corporate level, making the link between land management, farming waste and the impacts on other water users, giving some examples based in Scotland, and link this to the impact on the end consumer. They further explore aspects of individual social justice, considering the impact of debt and deprivation, and use statistics from England and Wales to support their arguments. Finally they consider the position of social justice from a global perspective, considering upland drainage systems and the impact that this has on coastal communities in the longer term.

We have debated some of the drivers for change and the impacts that these have on water availability, which creates local and global challenges in everything from supply of safe drinking water to improvements in land management and carbon reduction. In addition to these aspects of innovation, there are research scientists around the world looking at the longer term with a view to developing technology that will act as an enabler of sustainability.

Both Stuart Khan and Greg Lowry (with his co-author Matt Hotze) in the final two chapters present some new ideas and different approaches to dealing with water availability and removal of contamination, with an objective of maximising the use of recycled water. These techniques and processes will become more important in protecting our freshwater sources in the future. It is important that we continue to support investment in research and innovation if we are to achieve the objective of safe clean drinking water for the global population.

It is now worth considering policy as a driver for change in moving to a more sustainable water and healthier ecosystem. We can invest any amount of time and money in research, but if this is not supported by good policy then it is unlikely to realise full value and benefit. It is important then, that national governments agree an alignment of policy wherever possible and this should include legislative instruments and well as setting the right taxation levies.

As an example, in Scotland the government has created a Coordinated Agenda for Marine, Environment and Rural Affairs Science (CAMERAS, see www.scotland.gov.uk/Publications/2009/01/21125048/3) which links into the ecosystems services approach. The CAMERAS group is tasked with developing sustainable strategies for science research in the environmental sector and provides an exciting opportunity to align activities that link land management, marine and fresh water ecosystems. The strategies that are in place and under development are aligned to the United Kingdom and European research priorities. It is important that we take every opportunity in finding solutions together and recognise that many challenges cross a range of scientific and engineering disciplines.

Throughout this foreword we have only scratched the surface of the debate around creating a more sustainable approach to water management and we have touched upon some of the key influencing factors and the questions that they present us with.

What we can conclude from this is that water sustainability, climate change, ecological health, population changes, land management, urbanisation and economic prosperity are all intrinsically linked. It is therefore important to recognise that the challenges we face on a global scale can only be resolved by scientists, engineers and policy makers working together to create aligned objectives and strategies.

Richard Allan
Chief Scientist, Scottish Water

Preface

This book series focuses on issues in environmental science and technology which are of particular current interest and concern, as may be seen from the titles of other recent volumes which are listed on an earlier page. Few issues are of greater significance today than the subject of this book – the sustainability of water supplies to the growing populations throughout the world.

It is predicted that climate change will result in big changes to the global distribution of rainfall, causing drought and desertification in some regions but floods in others. Already there are signs of such change occurring, with particularly serious consequences for some of the poorer countries. The need for international cooperation in managing the effects of climate change and other influences on the hydrological cycle is becoming urgent: future wars may well be fought over water.

As is usual for books in this series, we have brought together a group of experts in the subject area to contribute articles covering a wide range of topics which bear on the overall theme of sustainable water. The authors of the seven chapters are drawn from institutions in the UK, Belgium, Germany, Ireland, Australia and the United States and apply their specialised knowledge to a wide range of relevant subjects, including policy making in the European Union, rural water supplies in Africa, chemical monitoring and analytical methods, water use in agriculture, social justice in supplying water, potable water recycling and sustainable water treatment. Although wide ranging, the coverage is not intended to be comprehensive, but the Foreword provides a context and also summarises the content.

The book aims to be useful not only to those in the water industry but more widely to policy makers and planners, researchers and environmental consultants, as well as to students in environmental science, technology, engineering and management courses. There is much here also to interest all concerned with major environmental issues such as climate change and the many other factors which influence the sustainability of water supplies.

Ronald E. Hester
Roy M. Harrison

Issues in Environmental Science and Technology, 31
Sustainable Water
Edited by R.E. Hester and R.M. Harrison
© Royal Society of Chemistry 2011
Published by the Royal Society of Chemistry, www.rsc.org

Contents

**Water Sustainability and Climate Change in the EU and Global Context –
Policy and Research Responses** **1**
Philippe Quevauviller

1	Introduction	2
2	Climate Change Impacts on Water	2
3	Policy Background	4
	3.1 Introduction	4
	3.2 EU Policies	5
	3.3 At International Level	12
4	Current Research	13
	4.1 Introduction	13
	4.2 Research into Climate Change Scenarios	15
	4.3 Research into Climate Change Impacts on the Water Environment and Cycle	16
	4.4 Research into Mitigation/Adaptation Options and Costs	17
	4.5 Research on Droughts and Water Scarcity	19
	4.6 Research on Floods	19
	4.7 Research Perspectives and Needs	20
5	Conclusions: Needs for Improving Science – Policy Links	21
	References	22

**Potential Impact of Climate Change on Improved and
Unimproved Water Supplies in Africa** **25**
Helen Bonsor, Alan MacDonald and Roger Calow

1	Introduction	26
2	Scenarios of Climate Change	28

Issues in Environmental Science and Technology, 31
Sustainable Water
Edited by R.E. Hester and R.M. Harrison
© Royal Society of Chemistry 2011
Published by the Royal Society of Chemistry, www.rsc.org

2.1   IPCC Fourth Assessment of Climate Change   28
2.2   Key Uncertainties in Climate Projections   29
      2.2.1   General Uncertainties in Climate
              Projections   29
      2.2.2   Uncertainties in African Climate
              Projections   30
2.3   Projected Climate Change in Africa   31
2.4   Climate Science since the IPCC Fourth
      Assessment: 4 °C Possibilities   32
2.5   Summary   34
3   Impacts of Climate Change on Rural Water Supply in
    Africa   34
3.1   A Framework for Discussion   34
3.2. Likely Impact of Climate Change on Available Water
      Resources   35
      3.2.1   General   35
      3.2.2   Surface Water Resources   35
      3.2.3   Groundwater Resources   37
3.3   Access to Reliable Water Supplies   39
3.4   Changing Water Demands   43
4   Summary   44
References   45

**The European Water Framework Directive – Chemical
Monitoring Programmes, Analytical Challenges and Results from an Irish
Case Study**   **50**
*Ulrich Borchers, David Schwesig, Ciaran O'Donnell and
Colman Concannon*

1   The Chemical Monitoring Approach of the WFD   51
1.1   Basic Principles and Approach   51
1.2   Environmental Quality Standards (EQS) and
      Resulting Monitoring Requirements   52
1.3   Design of Monitoring Programmes in the EU   54
      1.3.1   General   54
      1.3.2   Design of Surveillance and Operational Monitoring
              Programmes   54
      1.3.3   Sampling Strategy   55
      1.3.4   Sampling of Water and Suspended
              Particulate Matter (SPM)   55
1.4   Frequency of the Monitoring   56
2   Analytical Challenges of the WFD Monitoring   57
2.1   Analytical Methods for the Determination of Priority
      Substances in Water   57
2.2   The EU QA/QC Directive 2009/90/EC   58

   2.3 Priority Substances Difficult to Analyse   59
      2.3.1 Organochlorine Pesticides   59
      2.3.2 Polycyclic Aromatic Hydrocarbons (PAHs)   60
      2.3.3 Tributyltin Compounds   60
      2.3.4 Pentabromodiphenylether (PBDE)   62
      2.3.5 Short-Chain Chlorinated Paraffins (SCCPs)   62
3 Case Study: Surface Water Monitoring in Ireland   64
   3.1 Introduction   64
   3.2 Overview of Results of the Chemical Monitoring of Priority
       Substances   64
      3.2.1 Substances with Concentrations below LOQ   64
      3.2.2 Substances with Concentrations above LOQ
           (Positive Results)   68
      3.2.3 Substances with Concentrations above the EQS   68
   3.3 Discussion   72
      3.3.1 Mecoprop   72
      3.3.2 Glyphosate   73
      3.3.3 Polycyclic Aromatic Hydrocarbons (PAHs)   73
   3.4 Challenges and Pitfalls   74
      3.4.1 Tributyltin   74
      3.4.2 Di(2-ethylhexyl)phthalate (DEHP)   75
References   76

**Managing the Water Footprint of Irrigated Food Production
in England and Wales**   **78**
*Tim Hess, Jerry Knox, Melvyn Kay and
Keith Weatherhead*

1 Water Footprints – Understanding the Terminology   79
   1.1 Definition of Water Footprint   79
   1.2 "Blue" and "Green" Water   79
2 Water Use in Irrigated Agriculture   80
   2.1 Areas Irrigated and Volumes of Water Abstracted   80
   2.2 Irrigation Water Sources   82
   2.3 Location of Irrigation   82
3 Managing the Water Footprint   83
   3.1 Managing Water Better   83
      3.1.1 Improving Management to Increase Irrigation
           Efficiency   85
      3.1.2 Switching Technology to Increase Irrigation
           Application Uniformity   86
      3.1.3 Securing Water Resources and Using "Appropriate"
           Quality Water   87
   3.2 Managing Abstraction   87

4   Discussion   88
5   Conclusion   90
Acknowledgements   90
References   90

Social Justice and Water   **93**
Adrian McDonald, Martin Clarke, Peter Boden and David Kay

1   The Emergence of the Social Justice Concept   93
2   Definitions and Meaning   94
3   Water and Interaction with People   94
4   Water and Social Justice on a World Scale   95
5   Flooding and Social Justice   96
6   Water and Social Justice on a UK Scale   97
    6.1   What Price Water?   97
    6.2   Water and Social Justice in the UK at Fine Scale   99
          6.2.1   Why an Analysis of Water Debt: Water Debt and
                  Corporate Justice?   99
          6.2.2   Water Debt in Context   100
          6.2.3   The Linkage to Deprivation   102
    6.3   Developing a Socially just Response to Water Debt   107
7   Social Justice and Water Futures   109
References   111

**Safe Management of Chemical Contaminants for Planned Potable
Water Recycling**   **114**
Stuart Khan

1   Introduction: Planned Potable Water Recycling   115
2   Chemical Contaminants in Potable Water Recycling   116
3   Chemical Risk Assessment and Potable Water Recycling   118
4   Relative Risk   122
5   Direct Toxicity Testing   123
    5.1   *In vivo* Toxicity Testing   123
    5.2   *In vitro* Toxicity Testing   124
6   Indicator Chemicals and Surrogate Parameters   126
7   Probabilistic Water Treatment Performance Assessment   128
8   Australian Guidelines for Water Recycling   132
9   Conclusions   135
References   135

Nanotechnology for Sustainable Water Treatment **138**
Matt Hotze and Greg Lowry

1	Introduction	139
2	Disinfection and Oxidation Technologies	140
	2.1 Oligodynamic Processes	140
	2.2 Photo-Driven Processes	142
	2.2.1 Photocatalytic Semiconductors	142
	2.2.2 Fullerene Photosensitisation	143
3	Nanotechnology Improving Membranes for Water Treatment	145
	3.1 Nanocomposite Membranes	145
	3.2 Nanotube Embedded Membranes	147
	3.3 Monitoring Membrane Failure with Nanomaterials	148
4	Groundwater Remediation Using Nanotechnology	148
	4.1 Electrically Switched Ion Exchange (ESIX)	149
	4.2 Nano Reactive Zero Valent Iron Particles for *in situ* Groundwater Remediation	149
	4.3 Bimetallic Particles for Transformations	152
	4.4 Surface Modified Nanoparticles for *in situ* Groundwater Treatment	153
	4.4.1 Polymeric Surface Modification	154
	4.4.2 Emulsified NZVI	155
	4.4.3 NZVI Embedded onto Carriers or Supports	155
	4.4.4 Particles Embedded in Membranes	156
5	Sustainability Challenges	156
	5.1 Raw Materials	157
	5.2 Manufacturing	157
	5.3 Use and End of Life	158
6	Conclusions	159
	Acknowledgements	159
	References	159
	Subject Index	**165**

Editors

Ronald E. Hester, BSc, DSc (London), PhD (Cornell), FRSC, CChem

Ronald E. Hester is now Emeritus Professor of Chemistry in the University of York. He was for short periods a research fellow in Cambridge and an assistant professor at Cornell before being appointed to a lectureship in chemistry in York in 1965. He was a full professor in York from 1983 to 2001. His more than 300 publications are mainly in the area of vibrational spectroscopy, latterly focusing on time-resolved studies of photoreaction intermediates and on biomolecular systems in solution. He is active in environmental chemistry and is a founder member and former chairman of the Environment Group of the Royal Society of Chemistry and editor of 'Industry and the Environment in Perspective' (RSC, 1983) and 'Understanding Our Environment' (RSC, 1986). As a member of the Council of the UK Science and Engineering Research Council and several of its sub-committees, panels and boards, he has been heavily involved in national science policy and administration. He was, from 1991 to 1993, a member of the UK Department of the Environment Advisory Committee on Hazardous Substances and from 1995 to 2000 was a member of the Publications and Information Board of the Royal Society of Chemistry.

Roy M. Harrison, BSc, PhD, DSc (Birmingham), FRSC, CChem, FRMetS, Hon MFPH, Hon FFOM

Roy M. Harrison is Queen Elizabeth II Birmingham Centenary Professor of Environmental Health in the University of Birmingham. He was previously Lecturer in Environmental Sciences at the University of Lancaster and Reader and Director of the Institute of Aerosol Science at the University of Essex. His more than 300 publications are mainly in the field of environmental chemistry, although his current work includes studies of human health impacts of atmospheric pollutants as well as research into the chemistry of pollution phenomena. He is a past Chairman of the Environment Group of the Royal Society of Chemistry for whom he has edited 'Pollution: Causes, Effects and Control' (RSC, 1983; Fourth Edition, 2001), and 'Understanding our Environment: An Introduction to Environmental Chemistry and Pollution' (RSC, Third Edition, 1999). He has a close interest in scientific and policy aspects of air pollution, having been Chairman of the Department of Environment Quality of Urban Air Review Group and the DETR Atmospheric Particles Expert Group. He is currently a member of the DEFRA Air Quality Expert Group, the DEFRA Expert Panel on Air Quality Standards, and the Department of Health Committee on the Medical Effects of Air Pollutants.

List of Contributors

Peter Boden, *School of Geography, University of Leeds, LS2 9JT*

Helen Bonsor, *British Geological Survey, Murchison House, West Mains Rd, Edinburgh EH9 3LA, UK*

Ulrich Borchers, *IWW Water Centre , Moritzstrasse 26, D-45476 Muelheim an der Ruhr, Germany*

Roger Calow, *Overseas Development Institute, Westminster Bridge Road, London, UK, SE1 7JD*

Martin Clarke, *Department of Geography, Earth and Environment Faculty, University of Leeds, Leeds LS2 9JT, UK*

Colman Concannon, *EPA, Richview, Glenskeagh Road, Dublin 14, Ireland*

Tim Hess, *Department of Sustainable Systems, Cranfield University, Bedfordshire, MK43 0AL, UK*

Matt Hotze, Center for Environmental Implications of Nanotechnology (CEINT), Carnegie Mellon University, Civil and Environmental Engineering, 119 Porter Hall, Pittsburgh, PA 15213-3890, USA

David Kay, *Institute of Geography and Earth Science, Aberystwyth University, Aberystwyth, SA23 3DB, UK*

Melvyn Kay, *Department of Sustainable Systems, Cranfield University, Bedfordshire, MK43 0AL, UK*

Stuart Khan, *UNSW Water Research Centre, University of New South Wales, NSW, 2052, Australia*

Jerry Knox, *Department of Sustainable Systems, Cranfield University, Bedfordshire, MK43 0AL, UK*

Gregory Lowry, *Center for Environmental Implications of Nanotechnology (CEINT), Carnegie Mellon University, Civil and Environmental Engineering, 119 Porter Hall, Pittsburgh, PA 15213-3890, USA*

Alan MacDonald, *British Geological Survey, Murchison House, West Mains Rd, Edinburgh EH9 3LA, UK*

Adrian McDonald, *Department of Geography, Earth and Environment Faculty, University of Leeds, Leeds LS2 9JT, UK*

Ciaran O'Donnell, *EPA, Richview, Glenskeagh Road, Dublin 14, Ireland*

Philippe Quevauviller, *European Commission, DG Research, Brussels, Rue de la Loi 200, B-1049 Brussels, Belgium*

David Schwesig, *IWW Water Centre , Moritzstrasse 26, D-45476 Muelheim an der Ruhr, Germany*

Keith Weatherhead, *Department of Sustainable Systems, Cranfield University, Bedfordshire, MK43 0AL, UK*

Water Sustainability and Climate Change in the EU and Global Context – Policy and Research Responses

PHILIPPE QUEVAUVILLER*

ABSTRACT

Climate change impacts on the hydrological cycle (*e.g.* effects on atmospheric water vapour content, changes of precipitation patterns) have been linked to observed warming over several decades. Higher water temperatures and changes in extremes, including floods and droughts, are projected to affect water quality and exacerbate many forms of water pollution with possible negative impacts on ecosystems and human health, as well as water system reliability and operating costs. In addition, sea-level rise is projected to extend areas of salinisation of groundwater and estuaries, resulting in a decrease of freshwater availability for humans and ecosystems in coastal areas. Besides this, changes in water quantity and quality due to climate change are expected to affect food availability, water access and utilisation, especially in arid and semi-arid areas, as well as the operation of water infrastructure (*e.g.* hydropower, flood defences, irrigation systems). This chapter discusses how climate change might impact the reliability of current water management systems on the basis of expert reports prepared at global or EU level, namely reports of the Intergovernmental Panel on Climate Change (IPCC) and guidance documents of the Water Framework Directive Common Implementation Strategy. Examples of international research trends are described to illustrate on-going efforts to improve understanding and modelling of

---

*The views expressed in this chapter are purely those of the author and may not in any circumstances be regarded as stating an official position of the European Commission.

Issues in Environmental Science and Technology, 31
Sustainable Water
Edited by R.E. Hester and R.M. Harrison
© Royal Society of Chemistry 2011
Published by the Royal Society of Chemistry, www.rsc.org

climate changes related to the hydrological cycles at scales that are relevant to decision making (possibly linked to policy).

1 Introduction

According to the Technical Paper VI of the Intergovernmental Panel on Climate Change (IPCC),[1] observational records and climate projections provide abundant evidence that freshwater resources are vulnerable toward climate change, with wide-ranging consequences for human societies and ecosystems in Europe and worldwide. In particular, observed warming over several decades has been linked to changes in the large-scale hydrological cycle (*e.g.* effects on atmospheric water vapour content, changes of precipitation patterns with consequences on extreme floods and droughts). Higher water temperatures and changes in extremes, including floods and droughts, are projected to affect water quality and exacerbate many forms of water pollution from sediments, nutrients, dissolved organic carbon, pathogens, pesticides and salt, with possible negative impacts on ecosystems, human health, and water system reliability and operating costs.[1] In addition, sea-level rise is projected to extend areas of salinisation of groundwater and estuaries, resulting in a decrease of freshwater availability for humans and ecosystems in coastal areas. Besides this, changes in water quantity and quality due to climate change are expected to affect food availability, stability, access and utilisation, especially in arid and semi-arid areas, as well as the function and operation of water infrastructure (*e.g.* hydropower, flood defences and irrigation systems).

The consequences of climate change may alter the reliability of current water management systems. While quantitative projections of changes in precipitation, river flows and water levels at the river-basin scale remain uncertain, it is very likely that hydrological characteristics will change in the future.[1] Adaptation options are currently designed to ensure water supply during average and drought conditions, while mitigation measures are also developed to reduce the magnitude of impacts of global warming on water resources, in turn reducing adaptation needs (with, however, possible negative side effects such as, for example, increased water requirements for bio-energy crops, reforestation, *etc.*). The options to respond to climate change are closely linked to a range of policies covering different sectors, *e.g.* energy, health, food security, water and nature conservation. This requires that adaptation and mitigation measures are evaluated across multiple water-dependent sectors.

2 Climate Change Impacts on Water

There is far-reaching consensus among scientists that climate change is, at least to a certain extent, caused by human activities. According to the terminology of the EU Water Framework Directive (2000/60/EC) or WFD[2] discussed in section 3, direct climate change impacts on water resources should not be classified as an "anthropogenic pressure" in the narrow sense, since they cannot be mitigated by water managers' action. However, climate change impacts interact with and potentially aggravate other anthropogenic pressures and *could* therefore be considered as an anthropogenic pressure. For example, changes in precipitation and hotter/drier summer periods alter both the availability of water and the demand for water for uses such as agriculture. Lower water levels as a result of climate change may lead to an increase in the concentration of pollutants (less dilution). In addition, pressures on water from human activities may change as a result of climate change mitigation efforts,[3] *e.g.* targets for bioenergy production to reduce CO_2 emission from burning oil tend to increase pressures on water in several places while, on the other hand, the requirement of cleaner production techniques to reduce CO_2 emission might also support the development of more water-protective technologies.

With the change of rainfall patterns, seasonality and spatial distribution, impacts of climate change are reflected in influences on the quantity and quality of water resources and impacts on their uses, *e.g.* abstraction of both surface and ground waters.[3] Sustainable water resource management is hence closely connected to various drivers, including climate change, land cover and increasing water consumption (see Figure 1).

Long-term threats to groundwater resources could be linked to intense rainfalls, resulting in surface flooding rather than infiltration to groundwater. Water quality will also be affected in that run-off takes nutrients and pesticides from agricultural land and transfer them into rivers and lakes, for instance. Less availability of water resources will mean lower quality in some cases, *e.g.* droughts can have an impact on the ecology of rivers.[3] Extremes may also have

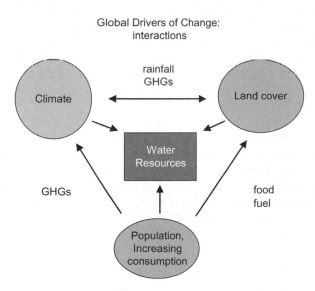

Figure 1 Sustainable water management in the context of global drivers of change. (Courtesy of R. Harding, WATCH project: *Water and Global Change*).

adverse effects on aquatic ecosystems, turning into not only a water shortage problem but also into a large environmental problem (up to desertification in some areas). Possible impacts to be anticipated also concern issues of water demand (increased demand and/or differing patterns), water infrastructures (*e.g.* in low-lying and coastal areas prone to flood risks, soil movement, effects on water treatment, *etc.*), sewer network operations (*e.g.* disruption due to extreme rainfalls with risks of pollution in industrial areas), threats to economic development of the water sector, *etc.*[3]

The assessment of climate change impacts on water resources implies a good knowledge of their global/regional distribution. The most sensitive hydrological systems should also be identified at the scale of river basins (this is linked to WFD river basin management principles, see section 3). This means that climate-induced changes in hydrological systems and processes should be better understood, in particular, variables such as river flows, groundwater and lake levels, soil moisture, evapotranspiration, snow cover, glacier extent, permafrost, *etc.*, as well as impacts on biodiversity.[3] Besides impacts on natural conditions, climate change impacts on sectoral water uses should also be evaluated, *e.g.* on agriculture (rain-fed and irrigated), forestry (including forest fires and deforestation), hydropower, navigation and water supply (domestic, agricultural and industrial). Furthermore, water-related impacts on infrastructure, health, transport, financial services (*e.g.* insurance sector), energy and tourism should also be reviewed. This should be done in an integrated way in order to tackle multi-risk evaluations at the river basin scale, distinguishing land-use changes due to human activities from climate-induced changes.

While climate change and climate change impacts research are progressing fast, there is still a lot of uncertainty, particularly with regard to water-related changes. There is also large uncertainty about future projections of climate change impacts on waters during the forthcoming decades. Over this timeframe, mean temperatures are expected to continue to rise but large year-to-year variations in precipitation probably will mask underlying regional trends for several decades. This implies that temperature-dependent processes (such as seasonal snowmelt, species' distribution and phenology, *etc.*) probably will manifest change in the first instance.[3] An increase of extreme events (floods and droughts) is also likely to occur. Uncertainties stem from different sources, *e.g.* difficulties in predicting future socio-economic development (scenario uncertainty), unsatisfactory model resolution and insufficient mathematical description of all global circulation processes (model uncertainty, especially for precipitation), lack of local hydrological localised models, *etc.* Attributing these hydrometeorological extremes to climate change is still uncertain because of a lack of accurate data and full scientific understanding of the functioning of the climate system.

3   Policy Background

3.1   Introduction

The need for policy responses to tackle climate change impacts on water is recognised worldwide. This is extensively expressed in the IPCC Technical

Paper on Water,[1] which is addressed primarily to policy-makers engaged in all areas related to freshwater resource management, climate change, strategic studies, spatial planning and socio-economic development. This Technical Paper evaluates the impacts of climate change on hydrological processes and regimes, and of freshwater resources (availability, quality, uses and management), at a worldwide scale, and highlights their implications for policy, looking at different sectors.[1] In particular, it provides recommendations regarding adaptation measures in regions prone to climate-change-related extremes about water resource management, ecosystems, agriculture and forestry, coastal systems, sanitation and human health. Some statements issued from the IPCC Technical Paper are summarised in Table 1.[1] This short introduction only serves to highlight that awareness for policy actions is growing worldwide but that, to date, no legal framework is in place to tackle climate change impacts on water at a global scale.

3.2 EU Policies

Compared with many international river basins worldwide which have no legally enforceable management framework, the situation in the European Union is developing towards a robust integrated water resources management system, with legal instruments being in place or in development. This section examines concrete policy steps that are either implemented or being developed in Europe. In the first place, EU water managers are currently implementing the Water Framework Directive (WFD),[2] which is the main legislative instrument for water protection. Details about operational policy measures related to this directive can be found in the literature.[4,5] The basic feature to be kept in mind in the context of this chapter is that the WFD is built upon the principles of river basin management planning (see Figure 2), considering all types of waters and pressures that may affect them, and designing programmes of measures (supported by extensive monitoring) to achieve "good status" objectives by 2015 (this concerns chemical and ecological status for surface waters, and chemical and quantitative status for groundwater). Climate change might affect and interact with all steps of WFD implementation, and thus on the status objectives, and this has been subject to in-depth discussions within the policy and scientific communities over the years 2008–2010 as reflected in the literature[3,6] and in a guidance document of the WFD Common Implementation Strategy examining river basin management in a changing climate.[7]

The WFD does not explicitly refer to risks posed by climate change to the achievement of environmental ("good status") objectives. However, several articles provide a framework to include climate change impacts into the planning process. In particular, the Annex II of the directive stipulates that *"Member States shall collect and maintain information on the type and magnitude of the significant anthropogenic pressures to which the surface water bodies in each river basin district are liable to be subject"*. With the far-reaching consensus that climate change is at least to a certain extent caused by human activities,[1] climate change impacts could fall into the category of "anthropogenic pressures".[3] It should be highlighted that these impacts cannot

Table 1   Summary of IPCC statements about climate change and water (after Bates *et al.*).[1]

Key statements	Additional comments
Observed warming over several decades has been linked to changes in the large-scale hydrological cycle.	This is reflected in increasing atmospheric water vapour content, changing precipitation patterns, intensity and extremes, reduced snow cover, and changes in soil moisture and runoff.
Climate model simulations for the 21st century are consistent in projecting precipitation increases in high latitudes (*very likely*) and parts of the topics, and decreases in some sub-tropical and lower mid-latitude regions (*likely*).	Outside these areas, the sign and magnitude of projected changes varies between models, leading to substantial uncertainty in precipitation projections. Thus, projections of future precipitation changes are more robust for some regions than for others. Projections become less consistent between models as spatial scale decreases.
By the middle of the 21st century, annual average river run-off and water availability are projected to increase as a result of climate change at high latitudes and in some wet tropical areas, and decrease over some dry regions at mid-latitudes and in the dry topics.	This statement excludes changes in non-climatic factors such as, for example, irrigation. Many semi-arid and arid areas (*e.g.* the Mediterranean Basin, Western USA, Southern Africa and North-Eastern Brazil) are particularly exposed to the impacts of climate change and are projected to suffer a decrease of water resources due to climate change (*high confidence*).
Increased precipitation intensity and variability are projected to increase the risks of flooding and drought in many areas.	The frequency of heavy precipitation events (or proportion of total rainfall from heavy falls) will be *very likely* to increase over most areas during the 21st century, with consequences for the risk of rain-generated floods. At the same time, the proportion of land surface in extreme drought is projected to increase (*likely*).
Water supplies stored in glaciers and snow cover are projected to decline in the course of the century.	This is linked to a projected reduction of water availability during warm and dry periods (through a seasonal shift in streamflow, an increase in the ratio of winter to annual flows, and reductions in low flows) in regions supplied by melt water from major mountain ranges, where more than one-sixth of the world's population currently live (*high confidence*).
Higher water temperatures and changes in extremes, including floods and droughts, are projected to affect quality and exacerbate many forms of water pollution.	Water pollution is projected to increase from sediments, nutrients, dissolved organic carbon, pathogens, pesticides and salt, as well as thermal pollution, with possible impacts on ecosystems, human health, and water system reliability and operating costs (*high*

Table 1   Continued.

Key statements	Additional comments
	confidence). In addition, sea-level rise is projected to extend areas of salinisation of groundwater and estuaries.
Globally, the negative impacts of future climate change on freshwater systems are expected to outweigh the benefits (*high confidence*).	By the 2050s, the area of land subject to increasing water stress due to climate change is projected to be more than double that with decreasing water stress. In many regions, the benefit linked to increased water supply is likely to be counterbalanced by the negative effects of increased precipitation variability and seasonal run-off shifts in water supply, water quality and flood risks (*high confidence*).
Changes in water quantity and quality due to climate change are expected to affect food availability, stability, access and utilisation.	This is expected to lead to decreased food security and increased vulnerability of poor rural farmers, especially in the arid and semi-arid tropics and Asian and African megadeltas.
Climate change affects the function and operation of existing water infrastructure (including hydropower, structural flood defences, drainage and irrigation systems), as well as water management practices.	Adverse effects of climate change on freshwater systems aggravate the impacts of other stresses, such as population growth, changing economic activity, land-use change and urbanisation.
Current water management practices may not be robust enough to cope with the impacts of climate change.	This may affect water supply reliability, flood risk, health, agriculture, energy and aquatic ecosystems. In many locations, water management cannot satisfactorily cope even with current climate variability, so that large flood and drought damages occur. Climatic and non-climatic factors, such as growth of population and damage potential, would exacerbate problems in the future (*high confidence*).
Climate change challenges the traditional assumption that past hydrological experience provides a good guide to future conditions.	The consequences of climate change may alter the reliability of current water management systems and water-related infrastructure. While quantitative projections of changes in precipitations, river flows and water levels at the river basin scale are uncertain, it is *very likely* that hydrological characteristics will change in the future.
Adaptation options designed to ensure water supply during average and drought conditions require integrated demand-side as well as supply-side strategies.	The former improve water-use efficiency, *e.g.* by recycling water. An expanded use of economic incentives, including metering and pricing, to encourage water conservation and development of water markets and implementation of virtual

Table 1 Continued.

Key statements	Additional comments
	water trade, holds considerable promise for water savings and the reallocation of water to highly efficient water uses.
Mitigation measures can reduce the magnitude of impacts of global warming on water resources, in turn reducing adaptation needs.	However, they can have considerable negative side effects, such as increased water requirements for afforestation/ reforestation activities or bio-energy crops. On the other hand, water management policy measures, *e.g.* hydro-dams, can influence greenhouse gas emissions.
Water resources management clearly impacts on many other policy areas.	Other areas concerned are, for example, energy, health, food security and nature conservation. Thus the appraisal of adaptation and mitigation options needs to be conducted across multiple water-dependent sectors.
Several gaps in knowledge exist in terms of observations and research needs related to climate change and water.	Observational data and data access are prerequisites for adaptive management, yet many observational networks are shrinking. There is a need to improve understanding and modelling of climate changes related to the hydrological cycle at scales relevant to decision-making.

be classified *sensu stricto* as anthropogenic pressure in the context of the WFD since they cannot be mitigated by water managers' actions. However, climate change might interact with and potentially exacerbate other anthropogenic pressures and should therefore be considered within the policy framework. For example, changes of precipitation patterns and drier summer periods might alter both the availability of water and water demands for agriculture and other uses, lower water levels might lead to pollution increases (less dilution), *etc*.

The climate sensitivity of the WFD has hence been studied in detail to evaluate the possible impacts on policy implementation.[3,7] The main considerations are summarised below:

- **Characterisation of water bodies and pressures.** In the context of the WFD, this involves a review of the impact of human activities (and related pressures) on the status of surface and groundwaters. Several factors used in this review are based on water system typologies that are themselves variables according to the climate (hence to climate change). This includes, for instance, river flow categories, energy of flow, precipitation patterns, water level fluctuations, *etc.*, with indirect impacts on pollution patterns (affecting both point and diffuse sources through changes of flows,

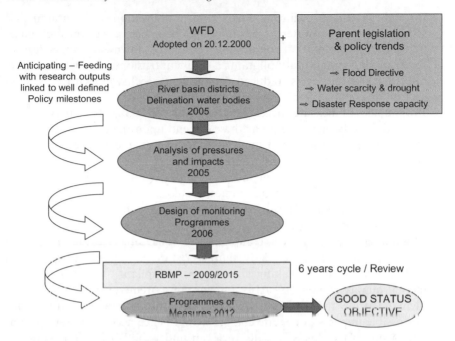

Figure 2 EU water policy milestones.

run-offs, *etc.*). This means that some characteristics of water bodies might be modified owing to climate change, with effects on their status.[3]

- **Risk assessment.** The characterisation of water bodies is an essential part of the WFD as it aims to forecast risks and calculate costs and benefits of the programme of measures. As expressed above, modifications of water bodies' characteristics due to climate change could lead to potential impacts on "good status" achievements (due to, for example, changes of water temperature, decreased dilution capacity of receiving waters, exceedence of water quality standards, changing metabolic rates of organisms, fish migration patterns, increased eutrophication, changes of river flows, *etc.* – the list is not exhaustive).[3]

- **Prevention of status deterioration.** Changes in the flow regime and physico-chemistry of rivers could have significant impacts on key species that could alter ecological status achievements, in particular in protected water bodies,[8] *e.g.* effects on spawning conditions for salmons, climate-driven shifts in species and community composition, *etc.*

- **Achievement of "good status".** Following on from the above, shifts in surface water bodies' characteristics might have effects on WFD status achievements,[3] in particular concerning compliance to environmental water quality standards (chemical status), impacts on fish mortality and biota composition (ecological status), *etc.* At the other end of the spectrum, increased flood frequency might also impact on status objectives through increased sediment loads and mobilisation of contaminated sediments.[9] Groundwater bodies may also be affected, *e.g.* through baseline shifting from natural conditions,[10] enhanced downward migration of, for example, agricultural pollutants,[11] saline intrusions in coastal aquifers due to rising sea levels, reduced groundwater recharge (with effects on quantitative status), *etc.*

- **Programmes of measures** necessary to deliver WFD objectives may also be affected directly or indirectly as these depend upon the above operational steps (characterisation, and analysis of pressures and impacts, in particular). The success of the programmes of measures will be closely related to the accurate characterisation in the first place, and flexibility to future changes in climate.[3] The programmes should also accommodate possible changes in behaviour ahead of climate change, such as adaptation measures in spatial strategies.

- **Monitoring** efficacy to check compliance to WFD objectives might also be affected by shifts in water body characteristics, *e.g.* increased river flows with greater dilution making sites more "compliant" towards environmental quality standards.[3] Also, impacts of extreme events may be problematic at low monitoring frequencies. Monitoring strategies in the light of possible impacts of climate change would hence need to be reviewed at regular intervals (this is actually foreseen under the WFD framework).

Opportunities given by the WFD river basin management planning for developing climate change adaptation policies has provided the reasoning for a guidance document which has been elaborated under the Common Implementation Strategy (CIS).[7] This guidance has been built upon principles of the European Commission's White Paper on "Adapting to climate change: towards a European framework for action",[12] which sets out a framework to reduce the EU's vulnerability to the impact of climate change. The White Paper is designed to complement Member States' actions (national and regional strategies) and support wider international adaptation efforts, particularly in developing countries, through cooperation with the United Nations Framework Convention on Climate Change (UNFCCC) towards a post-2012 climate agreement, which will address adaptation as well as mitigation. In particular, the White Paper identifies adaptation strategies to increase the resilience to climate change of a wide range of sectors, including by improving the management of water resources and ecosystems.

The Floods Directive[13] is included in a larger "Flood Action Programme" (see Figure 3) for the assessment and management of flood risks aimed at reducing the adverse consequences for human health, the environment, cultural heritage and economic activity associated with floods in Europe.

Besides the legislative framework, the package covers an optimal use of funding instruments (*e.g.* capacity-building projects), information exchange and research. This directive is coordinated with the implementation of the WFD from the second river basin management plan onward. It therefore provides a comprehensive mechanism for assessing and monitoring increased

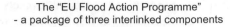

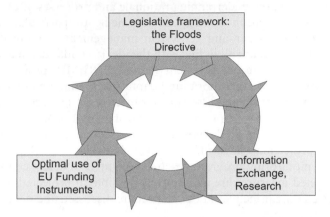

Figure 3   The EU Flood Action Programme.

risks of flooding due to climate change and for developing appropriate adaptation approaches. The coordinated approach with the river basin management plans will ensure an overall effective adaptation approach.

Alongside the increased risk of flooding, the policy implications of increased frequency of droughts due to climate change have to be considered. In this respect, the European Commission has adopted a Communication addressing the challenge of water scarcity and droughts in the EU;[14] this sets out a number of policy options for addressing the challenge of water scarcity. The Commission is conducting an annual European assessment of water scarcity and droughts, making it possible to monitor changes across Europe and to identify where further action is needed in response to climate change. In addition, a review of the strategy for water scarcity and droughts is planned for 2012.

Some recommended policy principles include the need to anticipate changes to water bodies that are climate sensitive, to understand the extent and causes of variability and change at reference sites (in particular, the most vulnerable ones), to assess direct and indirect influences on pressures due to climate change, and to identify and closely monitor climate change "hot spots".[7] Guiding principles for adaptation favour options that are robust to the uncertainties in climate projections and integrate cross-sectoral delivery of adaptation measures (in line with the river basin management planning).

Adaptation therefore has to deal with the protection of water supplies, maintaining services to consumers, protection of the environment and health, and limitation of financial impacts. Mitigation of the water industry's effects will also have to be addressed (regarding greenhouse gases emissions). Limitations to adaptation, however, exist in the sense that forecasts of changes will never be exact, and that adapting to climate change will in many cases be equivalent to preparing for a range of potential scenarios. Nonetheless, there are sufficient indications concerning potential impacts on relevant water management issues and trend changes to justify starting work on adaptation.[3] As noted previously, the underpinning rationale and processes of the WFD are consistent with adaptation strategy development. In particular, integrated approaches to land, water and ecosystem management, combined with the cyclical review process, are all in line with the ideals of adaptive management.[7] Steps in the River Basin Management Planning (RBMP) process provide a convenient structure for incorporating adaptation to climate change through risk appraisal, monitoring and assessment, objective setting, economic analysis and programmes of measures to achieve environmental objectives. As a minimum, Member States should demonstrate in the second and third cycle RBMP how climate projections have informed assessments of WFD pressures and impacts, how monitoring programmes are configured to detect climate change and how selected measures are as robust as possible to projected climate conditions. Key guiding principles in this respect are discussed and described in the CIS Guidance document.[7]

3.3 At International Level

Anticipating management options to tackle climate change in the world's river basins have been discussed in the literature,[15] giving examples of water-related climate change impacts likely to occur in populated basins in the world (more pronounced in basins impacted by dams than for basins with free-flowing rivers). While climate change impacts on water resources and proactive management efforts are recognised worldwide, there is not yet a "global policy" dealing with adaptation measures to climate change as regard water resources. This need is, however, clearly expressed by the UN Economic Commission for Europe (UNECE) in a guidance document published in 2009 (ref. 16). The key messages are along the same path as recommendations expressed in the IPCC Technical Paper on water,[1] in particular concerning the negative impacts on nearly all UNECE countries (though recognising that climate change might have positive consequences for some countries, such as prolonged growing season) linked to increased frequency and intensity of floods and droughts; worse water scarcity; intensified erosion and sedimentation; reduction in glacier and snow cover; sea level rise; salinisation; soil degradation and damage to water quality; ecosystems; and human health. The guidance states that implementing integrated water resources management will support adaptation to climate change (which goes along the conclusions of the CIS guidance document),[7] including planning at river basin level, strong intersectoral cooperation, public participation and making the best use of water resources.

Regarding policy, UNECE recommends that any policy needs to consider climate change as one of the many pressures on water resources (others include population growth, migration, globalisation, changing consumption patterns, and agricultural and industrial developments). Effective adaptation in this respect will require a cross-sectoral approach including at the trans-boundary level, in order to present possible conflicts between different sectors and consider trade-offs and synergies between adaptation and mitigation pressures. The guidance insists on the fact that legislation should not present barriers for adaptation and should be flexible enough to accommodate continuing environmental and socio-economic changes. Besides national legislation, a number of international agreements include provisions that can support the development of adaptation strategies (this includes the WFD, see above), and transboundary cooperation and policy might certainly build upon this basis.[16] Finally, the UNECE guidance stresses that uncertainty should not be a reason for inaction and highlights that action, knowledge and experience sharing, and research on adaptation should be pursued simultaneously and in a flexible way. This brings us to the next section which describes examples of research trends in the EU.

One of the key policy trends at international level is represented by the Hyogo Framework for Action, 2005–2015 (HFA). This programme was adopted by the United Nations in January 2005, with 168 nations committed to substantially reduce the loss of life and livelihoods from disasters. The scope of the HFA clearly goes beyond simply the water-related disasters, in particular extreme floods and droughts, as it includes, for example, earthquakes, tsunamis, volcanic eruptions and storms. The implementation of the HFA is under the responsibility of the United Nations International Strategy for Disaster Reduction (UN-ISDR) which is focal point in the UN system for the coordination of disaster reduction and for ensuring synergies among the disaster-reduction activities of the UN and regional organisations, and for activities in socio-economic and humanitarian fields. More information about the UN-ISDR objectives can be found in the 2010–2011 Biennial Work Programme.[17]

4 Current Research

4.1 Introduction

Research related to climate change and water is needed to improve understanding and modelling of climate changes related to the hydrological cycles at scales that are relevant to decision-making (possibly linked to policy). At present, scientific information about water-related impacts of climate change is not sufficient, especially with respect to water quality, aquatic ecosystems and groundwater, including their socio-economic dimensions. Research into climate-change impacts on the water cycle, and related extreme events (in particular floods and droughts), will help in improving the understanding and assessment of key drivers and their interactions, in order to better manage and mitigate risks and uncertainties. Arising questions concern scientific outcomes that are sufficiently mature to be taken aboard, policy development and which are the key research topics that need to be addressed at a European level.

Operational features related to the implementation of the WFD and, furthermore, those linked to climate change adaptation and mitigation, represent huge scientific challenges and research needs. These have been widely discussed in the scientific literature with regard to WFD implementation milestones,[3,4,18] *e.g.* regarding the long-term behaviour of rivers and transitional waters, eco-system responses, *etc.* Gaps in climate change research and water and related recommendations are also discussed and highlighted in IPCC,[1] UNECE[16] and UN-ISDR[19] documents. In addition, specific needs related to climate change and the WFD have been reviewed[18] and the European Environment Agency has also issued a Technical Report on climate change and water adaptation issues which clearly pinpoints research needs regarding policy support.[21] In particular, scientific knowledge needs to be strengthened or developed to advance our understanding of physical and biological function, process pathways, and pressures and risks posed by climate and land-use changes.[3] Research gaps also emerge at the interfaces between climate change and the natural processes of rivers, lakes and wetlands, including acidification, changes in mineralisation rates for nitrates, ecosystem modelling, *etc.*[20] Above all, fundamental understanding is lacking about how climate-change impacts on river flows, chemistry, sediment transport and hydromorphology will interact with ecology.[3] Further discussions about research gaps and the needs related to climate change and water can be found in the literature.[1,3,7,12,16,19–21]

Scientifically sound data and other information are essential for making climate projections while reducing their uncertainties, particularly for vulnerable groups and regions. This includes issues encompassing all aspects of the hydrological cycle, taking into consideration the needs of end-users, and including social and economic information. For instance, early warning systems are essential for preparedness for extreme weather events and should be developed at the trans-boundary level; they should also be closely linked to seasonal and long-term climate and weather forecast systems, as well as monitoring and observation systems.[16]

A large array of research activities currently focuses on water–climate interactions and impacts of climate change on the water cycle. As an example, climate–water-cycle feedbacks in Southern Europe have been widely studied, contributing to the understanding of the role of land-use in the climate–water interactions.[22] Other examples concern potential impacts of a warming climate on water availability in snow-dominated regions, particularly the consequences for the hydrological cycle and where water supply is dominated by melting snow or ice.[23] Many more research trends are described elsewhere,[3,6,7] some being detailed in the following sections.

Research on climate change is often closely linked to policy developments at EU level, as highlighted in the White Paper on *Adaptation to Climate Change*,[12] and on-going discussions about integration of adaptation and mitigation measures in the river basin management planning of the Water Framework Directive. Scientific outputs are also contributing to international policies and debates, in particular through inputs to IPCC assessment reports and UNFCCC documents.

More specifically, the European Commission is funding research through its Framework Programme for Research and Technological Development. In this context, projects of the 6th Framework Programme (2002–2006) [in particular, projects funded under the "Global Change and Ecosystems" sub-priority] and

FP7 - Environment (incl. Climate Change) 2007-2013 / 1890 Mio Euro

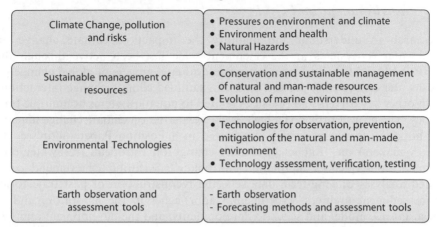

Figure 4 The research areas of the FP7 Environment Theme.

of the on-going 7th Framework Programme (2007–2013) or FP7 [in particular, projects funded under the "Environment (including climate change)" theme] largely contribute to gathering knowledge relevant to climate change adaptation in the context of the WFD river-basin management planning. Research areas covered by the FP7 Environment theme are summarised in Figure 4; they are exemplified by projects described below (the list is obviously far from being exhaustive – an updated list of projects in support of climate change research is available),[24] highlighting their potential to be linked to policy developments.

4.2   Research into Climate Change Scenarios

Research on climate change scenarios and predictions has been ongoing and expanding in the last few decades. For example, the PRUDENCE (Prediction of Regional scenarios and Uncertainties for Defining EuropeaN Climate change risks and Effects) project (2001–2004)[25] has provided a series of high-resolution climate change scenarios for 2071–2100, including an analysis of the variability and level of confidence in these scenarios as a function of uncertainties in model formulation, natural/internal climate variability and alternative scenarios of future atmospheric composition. A continuation of this research line is illustrated by the ENSEMBLES (based on predictions of climate changes and their impacts) project (2004–2009),[26] which integrates climate change impact studies into an ensemble prediction system, quantifies the uncertainty in long-term predictions of climate change and provides a reliable quantitative risk assessment of long-term climate change and its impacts. It includes the production of Regional Climate Scenarios for Impact Assessments and the formulation of very high resolution Regional Climate Model Ensembles for Europe.

4.3 Research into Climate Change Impacts on the Water Environment and Cycle

Research to understand and quantify the impact of climate change on freshwater ecosystems at the catchment scale has been active through the EURO-LIMPACS (European project to evaluate impacts of global change on freshwater ecosystems) project,[27] which examined climate change interactions with other key drivers and pressures related to aquatic systems at multiple time scales. The project provided a high level of expertise on climate change impacts on aquatic ecosystems, which is reflected in a Position Paper (addressed to policy-makers) on "Impact of climate change on European freshwater ecosystems: consequences, adaptation and policy". Scientific achievements combined analyses of long-term data sets, the reconstruction of past trajectories from sediment archives, experimental approaches in the laboratory and in mesoscosms, model and scenario developments, and the development and test of Decision Support Systems (DSS). The results from this research are expected to assist in: (i) assessing the potential impacts of global change at local-to-regional scales on different water bodies (excluding groundwaters and coastal waters) across the wide range of European climates, geomorphology types, land use, and human impact; (ii) developing a unified system of ecosystem health indicators; and (iii) reviewing the multiple restoration strategies and the impact of climate change on those concerning key issues such as hydrological and thermal changes, habitat degradation, eutrophication, acidification, and long-range atmospheric transfer of pollutants.

Specific research on climate change impacts on the global water cycle is carried out under the WATCH (global change and water) project[28] which unites different expertises (hydrologists, climatologists and water use experts) to examine the components of the current and future global water cycles, evaluate their uncertainties and clarify the overall vulnerability of global water resources related to the main societal and economic sectors. The project is developing a number of global and regional datasets to facilitate the assessment of changes in the water cycle, including case studies in river basins located in the EU. In parallel, a conceptual modelling framework is being developed to provide consistent modelling results and transfer information between scientists and stakeholders. This will include methodologies to handle biases in climate model output and quantify the resulting uncertainties in estimates of future components of the global water cycle. WATCH aims to increase our understanding of drought and large-scale flood development for the past and future climates through studies at different scales (global, regional, river basin). Current generation of large-scale models (*e.g.* climate models and global hydrological models) are tested against more detailed (hydrological) models to explore their ability to predict droughts and floods. Five test basins within Europe are being used to translate water resources applications from the global water cycle system to river basins.

The assessment of climate change impacts on water resources is also being studied in focused aquatic environment, *e.g.* the Mediterranean area through the CIRCE (climate change and impact research: the Mediterranean environment) project.[29] In particular, research is carried out to investigate how strongly climate variations induce significant changes in the hydrological cycle, *e.g.* increasing atmospheric water vapour, changing precipitation patterns and intensity, and changes in soil moisture and run-off. The project collects data from observations to quantify those changes and to develop a regional climate model able to analyse the conditions in the Mediterranean area. The investigations concern surface water, groundwater, coastal aquifers and the interactions between them. Both water quantity and quality issues are taken into account. The final goal of this project is to produce an assessment (RACCM – Regional Assessment of Climate Change in the Mediterranean) to be used to deepen the understanding of the impact of climate change on water resources and to suggest potential adaptation measures.

A more focused research is reflected by the on-going ACQWA (Assessing Climate change impacts on the Quantity and quality of WAter) project,[30] which investigates the consequences of climate change in mountain regions where snow and ice is currently an important part of the hydrological cycle. Numerical models are used to predict shifts in water amount by 2050, and how these changes will impact upon socio-economic sectors such as energy, tourism and agriculture. There will be focused studies on governance issues and ways of alleviating possible conflicts of interests between economic actors competing for dwindling water resources. Following a first phase of research in the data-rich European Alps, the models and methods will be applied to non-European regions such as the Andes and the Central Asian mountains, where climatic change and changing snow, ice and water resources will be a source of concern but also of opportunity in the future.

A related project (also dealing with climate change impacts on glaciers) has been launched in 2009: the HighNoon (adaptation to changing water resources availability in Northern India with respect to Himalayan glacier retreat and changing monsoon pattern) project will assess the impact of Himalayan glacier retreat, explore possible changes of the Indian summer monsoon on water resources in Northern India, and recommend appropriate and efficient response strategies for adapting to hydrological extreme events such as floods and droughts.[31] This project illustrates international cooperation efforts in climate change research which are also illustrated by other projects, *e.g.* a recently launched cluster on "Climate change impacts on water and security" which builds up cooperation among EU countries and neighbouring Mediterranean countries.[32]

4.4   *Research into Mitigation/Adaptation Options and Costs*

Mitigation/adaptation options to respond to climate change conditions have been developed, tested and evaluated within the AquaStress (mitigation of water stress through new approaches to integrating management, technical, economic and institutional instruments) integrated project,[33] leading to the definition of mitigation options exploiting new interfaces between technologies and social approaches, as well as economical and institutional settings. Particular emphasis has been given to methods, tools and guidelines, *e.g.* for groundwater modeling, groundwater recharge, improved crop policies, to facilitate a holistic approach to manage water supply and water demand. Several lessons can be derived from the AquaStress experience on improved approaches to integrated and participative water management, which is considered fundamental for adaptation to changing conditions.

Adaptation and mitigation strategies in support of European Climate Policy have also been investigated within the framework of the ADAM (assessing adaptation and mitigation policies) project[34] which developed long-term policy options/scenarios that could contribute to the EU's 2 °C target and targets for adaptation. The project made significant contributions to climate change policy developments through regular policy briefs, highlighting that Green House Gas emissions could be technically reduced in Europe by up to 80% by 2050. This is obviously only indirectly linked to river basin management developments but nevertheless it has consequences for the way integrated water resource management will have to evolve over the forthcoming decades.

Increasing uncertainties due to the accelerating pace and greater dimension of changes (*e.g.* climatic and demographic changes) and their impact on water resource management have been investigated by the NeWater (adaptive integrated water resources management) integrated project.[35] The project studied challenges such as climate change, flood-plain management and endangered ecosystems in order to move from the current regimes of river basin water management to more adaptive regimes in the future. In other words, NeWater aimed to understand and facilitate change towards more adaptive processes of Integrated Water Resources Management. Seven river basins (Amu Darya, Elbe, Guadiana, Nile, Orange, Rhine and Tisza) were selected as case study areas to establish the link between practical activities and advances in thematic research and tool development. The project has produced a book on *Climate Change Adaptation in the Water Sector* and twelve publicly available synthesis products which are of direct interest to policy implementation and development (including databases, guidelines on uncertainty in adaptive management, the climate and water adaptation book with various adaptation strategies, evaluation of water resources scenarios in the case studies, a guidebook on adaptive water management, *etc.*). All the reports and tools are available on the project webpage.

Besides the development of mitigation/adaptation strategies, an important element is the economic valuation of identified measures. In this respect, research has contributed to develop scenarios and quantify environmental and resource costs and benefits linked to adaptation to climate change within the framework of the AQUAMONEY (assessment of the environmental and resource costs and benefits of water services) project.[36] Efforts are being pursued with the recently launched ClimateCost (full costs of climate change) project,[37] which builds up on results of AQUAMONEY and ADAM to further develop climate change and socio-economic scenarios with a quantification of related costs, including an assessment of the physical effects and economic damages of major catastrophic events.

Finally, specific inputs for the identification of gaps that would have possible effects on the WFD implementation in combating climate impacts on water are being studied by the ClimateWater (bridging the gaps between adaptation strategies of climate change impacts and European water policies) project.[38] This project will identify the adaptation strategies that were developed in Europe and globally for dealing with climate change impacts on water resources and aquatic ecosystems (preventing, eliminating, combating and mitigating).

4.5 Research on Droughts and Water Scarcity

Besides research on management options addressed by AquaStress (see section 4.4), specific research needs on droughts are being discussed in the XEROCHORE (an exercise to assess research needs and policy choices in areas of drought) Support Action.[39] This Support Action is currently establishing the state of the art of drought-related national and regional policies and plans, and will lay down a roadmap that will identify research gaps on various drought aspects (climate, hydrology, impacts, management and policy) and the steps to be taken in order to fill them. In particular, support for European Drought Policy will be provided through expert recommendations about impact assessment, policy-making, drought in the context of integrated water resources management and guidance on appropriate responses for stakeholders. The large consortium (over 80 organisations) is closely linked to the European Drought Centre and the CIS Working Group on Water Scarcity and Drought, which has basically led to the development of an internationally recognised exchange platform on drought issues between the research and policy communities. This is strengthened by links established with relevant Research and Technological Development (RTD) projects, which include drought components, *e.g.* WATCH, CIRCE, as well as the recently launched MIRAGE (Mediterranean intermittent river management) project on Intermittent River Management.[40] It is expected that the exchange platform, now established and developed within the XEROCHORE project, will be further strengthened by the European Commission through the clustering of projects dealing with climate change and water security (including drought aspects) from 2010 onward.

4.6 Research on Floods

The project most relevant to flood research carried out within the years 2004–2009 at EU level in support of the Flood Directive is certainly the FLOODsite (integrated flood risk analysis and management methodologies) Integrated Project.[41] The project was interdisciplinary, integrating expertise from across the environmental and social sciences, as well as technology, spatial planning and management. The notion of "integrated flood" risk management now tends towards a change of policy from one of flood defence, to flood risks being managed but not eliminated. The project has developed robust methods of flood risk assessment and management, and decision support systems which have been largely tested in pilot sites. Regular contacts with the European Working Group on Floods of the WFD Common Implementation Strategy have enabled the policy community to be informed about progress on flood risk management. More than 100 research reports are available for public download on the project website.

Flash flood events and predictive scenarios have been studied by the FLASH (observations, analysis and modelling of lightning activity in thunderstorms for use in short-term forecasting of flash floods) project[42] on the basis of the collection and analysis of lightning data and precipitation observations. Research is continuing on how to reduce loss of life and economic damage, through the improvement of the preparedness and the operational risk management for flash floods and debris flow-generating events, as currently undertaken by the IMPRINTS (improving preparedness and risk management for flash floods and debris flow events) project,[43] which also studies how to contribute to sustainable development through reducing damage to the environment. The project will produce methods and tools to be used by emergency agencies and utility companies responsible for the management of these extreme events and associated effects. Impacts of future changes, including climatic, land use and socio-economic, will be analysed in order to provide guidelines for mitigation and adaptation measures.

Finally, the recently launched project CORFU (collaborative research on flood resilience in urban areas) will look at advanced and novel strategies and provide adequate measures for improved flood management in cities, focusing on Europe-Asia cooperation.[44] The differences in urban flooding problems in Asia and in Europe range from levels of economic development, infrastructure age, social systems and decision-making processes, to prevailing drainage methods, seasonality of rainfall patterns and climate change trends. The project will use these differences to create synergies that will bring an improved quality to flood management strategies globally.

4.7   Research Perspectives and Needs

Modelling capabilities need to be improved, and appropriate tools need to be developed to advance the ability for assessing climate effects on water resources and uses. New research areas (resulting from the 2009 call for proposals of the 7th EU Framework Programme) will investigate novel observation methods/ techniques and modelling, and socio-economic factor analyses to reduce existing uncertainties in climate change impact analysis and to create an integrated quantitative risk and vulnerability assessment tool. In particular, impacts on key strategic sectors, such as agriculture and tourism, will be investigated as well as macro-economic implications of water availability in terms of regional income, consumption, investment, trade flows, industrial structure and competitiveness, with focus on Southern Europe, North Africa and the Middle East.

In terms of perspectives, research should look into the evaluation of climate change adaptation and mitigation measures across multiple water-dependent sectors, and investigate interactions among water and other environmental compartments (sediment, soil and air). At present, scientific information about water-related impacts of climate change is not sufficient, especially with respect to translation of climate model output to the river-basin scale (matching the scale of the WFD River Basin Management Plans), water quality, aquatic ecosystems and groundwater, including their socio-economic dimensions. Research into climate change impacts on the water cycle from the global to the regional and river-basin scale is essential, to improve the understanding and assessment of key drivers and their interactions, in order to better manage and mitigate risks affecting the water cycle and to reduce uncertainties in policy responses. This also includes research related to disaster risk reduction, to improve understanding and modelling of extreme events related to the hydrological cycles at scales that are relevant to decision making (possibly linked to policy).

5   Conclusions: Needs for Improving Science – Policy Links

The need to act to protect water resources and cope with water stress and other effects of climate change is now well recognised worldwide. The integrated water resources management (IWRM) principle is increasingly used as a basis to manage water resources effectively at river basin and trans-boundary levels. This principle is made operational in Europe through the Water Framework Directive, with the aim to achieve "good status" for all waters, reduce water consumption, create the resources for maintenance and upgrading of water infrastructure through the user pays principle, and ensure that all stakeholders are consulted and given a voice.

As a concluding remark, it should be recalled that policy orientations rely on scientific evidence. In this respect, the efficient use of science represents an increasing challenge for the scientific and policy-making community, the private sector, NGOs, citizens' associations and professional organisations. The need to improve the role that science plays in environmental policy-making has been widely debated over the last few years. In particular, the need has been identified to ensure better linkages between policy needs and research programmes, with enhanced coordination regarding programme planning, project selection and management, and mechanisms for knowledge transfer to ensure that outputs from research programmes really do contribute to policy development, implementation and review. This issue has been discussed in depth in the water sector at European Union level for more than five years, underlining the need to develop a conceptual framework for a science–policy interface related to water which would gather together various initiatives and knowledge. This is discussed in depth in a recent book[45] and is illustrated in Figure 5, showing the necessary links between research recommendations or tools and "users" (policy-makers, stakeholders, water managers) and the need to ensure a "memory" of scientific information (facilitated by various dedicated websites), demonstration of the applicability of the research and dissemination through

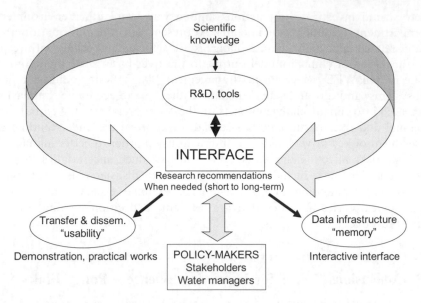

Figure 5 Needs of interface between science and policy (adapted from Quevauviller).[45]

appropriate communication and "translation" of the scientific information. This closes the present chapter in underlining that the bridge between policy and research is not trivial and deserves more attention from all actors concerned.

References

1. B. C. Bates, Z. W. Kundzewicz, S. Wu and J. P. Palutikof, *Climate Change and Water*, Technical Paper of the Intergovernmental Panel on Climate Change, IPCC Secretariat, Geneva, 2008, pp. 210.
2. European Commission, Directive 2000/60/EC of the European Parliament and of the Council of 23 October 2000 establishing a framework for Community action in the field of water policy, *Off. J. Eur. Communities*, **L 327**, 22.12.2000, 1.
3. R. L. Wilby, H. G. Orr, M. Hedger, D. Forrow and M. Blackmore, *Environ. Int.*, 2006, **32**, 1043.
4. P. Quevauviller, U. Borchers, K. C. Thompson and T. Simonart, *The Water Framework Directive – Ecological and Chemical Status Monitoring*, John Wiley and Sons Ltd., Chichester, UK, 2008.
5. P. Chave, *The EU Water Framework Directive*, IWA Publishing, London, UK, 2007.
6. F. Ludwig, P. Kabat, H. van Schaik and M. van der Valk, *Climate Change Adaptation in the Water Sector*, Earthscan, London, 2009.

7. European Commission, *River Basin Management in a Changing Climate*, Common Implementation Strategy for the Water Framework Directive, Guidance Document No. 24, ISBN 978-92-79-14298-7, 2009.
8. K. J. Limbrick, P. G. Withehead, D. Butterfield and N. Reynard, *Sci. Total Environ.*, 2000, **251–252**, 539.
9. R. L. Wilby, H. Y. Dalgleish and I. D. L. Foster, *Earth Surf. Process Landforms*, 1997, **22**, 353.
10. W. M. Edmunds and P. Shand, *Natural Groundwater Quality*, Blackwell Publishing, Oxford, 2008.
11. P. Stålnacke, A. Grimvall, C. Libiseller, M. Laznik and I. Kokorite, *J. Hydrol.*, 2003, **283**, 184.
12. European Commission, White Paper; Adapting to Climate change: Towards a European Framework for Action, COM(2009), 147 final, 2009.
13. European Commission, Directive 2007/60/EC of the European Parliament and of the Council of 23 October 2007 on the assessment and management of flood risks, *Off. J. Eur. Communities*, **L 288**, 6.11.2007, p. 27.
14. European Commission, Communication to the European Parliament and the Council – Addressing the challenge of water scarcity and droughts in the European Union, COM/2007/04141 final, 2007.
15. M. A. Palmer, C. A. Reidy Liermann, C. Nilsson, M. Florke, J. Alcamo, P. S. Lake and N. Bond, *Frontiers Ecol. Environ.*, 2008, **6**.
16. United Nations, *Guidance on Water and Adaptation to Climate Change*, United Nations, New York and Geneva, ISBN: 978-92-1-117010-8, 2009.
17. United Nations, *2010–2011 Biennial Work Programme*, International Strategy for Disaster Reduction (ISDR), Geneva, 2009.
18. P. Quevauviller, C. Fragakis and P. Balabanis, in *Water Systems Science and Policy Interfacing*, P. Quevauviller (ed), RSC Publishing, Cambridge, 2009, ch. 1.4, pp. 52.
19. United Nations, *Reducing Disaster Risks through Science: Issues and Actions*, ISDR Scientific and Technical Committee, 2009, Geneva.
20. R. L. Wilby, M. M. Hedger and C. Parker (eds), *Perspectives on Climate Change Science*, Environment Agency, Bristol, 2004, pp. 42.
21. European Environment Agency, *Climate Change and Water Adaptation Issues*, Technical Report No. 2/2007, 2007, Copenhagen.
22. M. M. Millán, M. J. Estrela, M. J. Sanz and E. Mantilla, *J. Climate*, 2005, **18**, 684.
23. T. P. Barnett, J. C. Adam and D. P. Lettenmaier, *Nature*, 2005, **438**, 303.
24. European Commission, *European Research Framework Programme: Research on Climate Change*, European Commission, EUR 23609, 2009.
25. PRUDENCE, Prediction of Regional scenarios and Uncertainties for Defining EuropeaN Climate change risks and Effects; http://prudence.dmi.dk/et
26. ENSEMBLES, based Predictions of Climate Changes and their Impacts; http://www.ensembles-eu.org/
27. EUROLIMPACS, European project to evaluate impacts of global change on freshwater ecosystems; http://www.eurolimpacs.ucl.ac.uk

28. WATCH, Global change and water; www.eu-watch.org
29. CIRCE, Climate change and impact research: the Mediterranean environment; http://www.circeproject.eu
30. ACQWA, Assessing Climate change impacts on the Quantity and quality of Water; www.acqwa.ch
31. HighNoon, Adaptation to changing water resources availability in Northern India with respect to Himalayan glacier retreat and changing monsoon pattern; http://www.eu-highnoon.org
32. R. Roson, Climate change impacts on water and security, cluster of CLIMB/WASSERMed/CLICO projects, 7th Framework Programme, in *International Conference on Integrated River Basin Management under the WFD*, Lille, 26–28 April 2010.
33. AQUASTRESS, Mitigation of water stress through new approaches to integrating management, technical, economic and institutional instruments; http://www.aquastress.net.
34. ADAM, Assessing adaptation and mitigation policies; www.adamproject.eu
35. NEWATER, Adaptive integrated water resources management; www.newater.info
36. AQUAMONEY, Assessment of the environmental and resource costs and benefits of Water services; http://www.aquamoney.ecologic-events.de/.
37. ClimateCost, Full costs of climate change; http://www.climatecost.cc/ClimateCost/Welcome.html
38. ClimateWater, Bridging the gaps between adaptation strategies of climate change impacts and European water policies; http://www.climatewater.org/
39. XEROCHORE, An exercise to assess research needs and policy choices in areas of drought; http://www.feem-project.net/xerochore/
40. MIRAGE, Mediterranean intermittent river management; http://www.mirage-project.eu/index.php
41. FLOODSite, Integrated flood risk analysis and management methodologies; www.floodsite.net
42. FLASH, Observations, analysis and modeling of lightning activity in thunderstorms, for use in short term forecasting of flash Floods; www.flashproject.org
43. IMPRINTS, Improving preparedness and risk management for flash floods and Debris flow events; http://www.imprints-fp7.eu/
44. CORFU, Collaborative research on flood resilience in urban areas; http://lib.bioinfo.pl/projects/view/11063
45. P. Quevauviller, *Water Systems Science and Policy Interfacing*, RSC Publishing, Cambridge, ISBN: 978–1–84755–861–9, 2010, pp. 430.

Potential Impact of Climate Change on Improved and Unimproved Water Supplies in Africa

HELEN BONSOR*, ALAN MACDONALD AND ROGER CALOW

ABSTRACT

With significant climate change predicted in Africa over the next century, this chapter explores a key question: *how will rural water supplies in Africa be affected*? Approximately 550 million people in Africa live in rural communities and are reliant on water resources within walking distance of their community for drinking water. Less than half have access to *improved* sources (generally large diameter wells, springs or boreholes equipped with hand pumps); the majority rely on *unimproved* sources, such as open water and shallow wells. Major climate modelling uncertainties, combined with rapid socio-economic change, make predicting the future state of African water resources difficult; an appropriate response to climate change is to assume much greater uncertainty in climate and intensification of past climate variability. Based on this assumption the following should be considered:

1. Those relying on unimproved water sources (300 million in rural Africa) are likely to be most affected by climate change because unimproved sources often use highly vulnerable water resources.
2. Improved rural water supplies in Africa are overwhelmingly dependent on groundwater, due to the unreliability of other sources.
3. Climate change is unlikely to lead to a continent-wide failure of improved rural water sources that access deeper groundwater (generally

---

*Corresponding author.

Issues in Environmental Science and Technology, 31
Sustainable Water
Edited by R.E. Hester and R.M. Harrison
© Royal Society of Chemistry 2011
Published by the Royal Society of Chemistry, www.rsc.org

over 20 metres below ground surface) through boreholes or deep wells. This is because groundwater-based domestic supply requires little recharge, and the groundwater resources at depth will generally be of sufficient storage capacity to remain a secure water resource. However, a significant minority of people could be affected if the frequency and length of drought increases – particularly those in areas with limited groundwater storage.

4. In most areas, the key determinants of water security will continue to be driven by access to water rather than absolute water availability. Extending access, and ensuring that targeting and technology decisions are informed by an understanding of groundwater conditions, will become increasingly important.

5. Accelerating groundwater development for irrigation could increase food production, raise farm incomes and reduce overall vulnerability. However, *ad hoc* development could threaten domestic supplies and, in some areas, lead to groundwater depletion.

Although climate change will undoubtedly be important in determining future water security, other drivers (such as population growth and rising food demands) are likely to provide greater pressure on rural water supplies.

1   Introduction

There is growing concern that climate change will have greatest impact on the world's poorest and most vulnerable people. The poor often have few options of where to access food and water and are highly reliant on the natural resources found close to their home and community. One group perceived to be highly vulnerable are the rural poor in Africa. With climate predicted to change significantly over the continent in the next century, this chapter explores the key question: *how will rural water supplies be affected?*

There are few large-scale public water supplies in rural Africa and, as a result, rural communities (some 550 million people in total) are generally reliant on water resources found within walking distance of their homes. The Joint Monitoring Programme[1] of WHO and UNICEF identifies three categories of drinking water supply: (a) water *piped* into the dwelling, plot or yard; (b) *other improved* sources (including public taps, protected springs, handpumps and rainwater harvesting); and (c) *unimproved* sources (open water, unprotected from contamination). For rural sub-Saharan Africa, 5% of the population have piped water, 41% are reliant on other improved sources (mostly wells and boreholes equipped with handpumps) and 54% (300 million people) have to rely on unimproved sources.[1]

Figure 1 shows examples of the improved water sources typically found across Africa. These commonly exploit groundwater resources through boreholes equipped with a handpump, large diameter wells, or springs.[2]

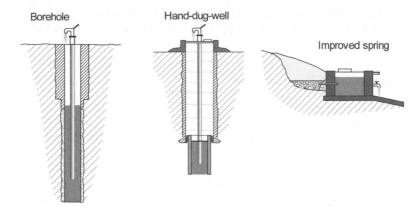

Figure 1 The most common types of improved rural water supplies within Africa.[2]

Figure 2 Unimproved sources are commonly shallow traditional wells (usually less than 10 m deep) within thick soils (left), or unprotected surface ponds and seepages (right).

Unimproved sources (on which the majority rely) comprise ponds, small dugouts, streams and commonly hand-dug wells less than 10 m deep (see Figure 2). These unimproved supplies often fail in the dry season, resulting in people having to walk longer distances to collect drinking water from more sustainable sources.[3]

The rise in temperature predicted with climate change will affect water resources and, therefore, all drinking water supplies – but the prediction of impacts is both complex and uncertain.[4] With such uncertainty there is a high risk that policies are introduced and practices adopted which are inappropriate and which could actually reduce water security.[3] To help reduce uncertainty, experience of water usage and water source behaviour during droughts in Africa is used to help understand the likely impacts of a changing climate.[5] These interdisciplinary studies indicate that water security is determined by three factors: absolute water resource availability, access to these water resources and changing water demand and use. By understanding the interplay between these three factors, more reliable predictions can be made and policies developed which are more likely to reflect best options in future decades.

2   Scenarios of Climate Change

Warming of the climate, by whatever means, is now unequivocal, with recorded increases since 1960 of global mean air temperature (*ca.* $+1\,°C$), sea surface temperature (*ca.* $+1\,°C$) and sea level rise (*ca.* $+150\,mm$).[4] With continued population growth and increased global economic development, climate change is predicted to occur at a much faster rate over the next 50 years and there is a growing body of evidence that a global mean temperature rise of $2.4\,°C$ will occur by 2100 regardless of any future emissions cuts.[6,7] Within Africa, such climate change is likely to cause a $3–4\,°C$ rise in land surface temperature and an intensification of the existing climatic and hydrological variability.[4,8]

2.1   IPCC Fourth Assessment of Climate Change

The Intergovernmental Panel on Climate Change (IPCC) has, in the last 20 years, published several reports synthesising climate change science and the Fourth Assessment (AR4), published in 2007, still provides the most comprehensive overview of likely climate change to date. Climate change is predicted by the IPCC based on a series of different emissions scenarios resulting from various lines of global socio-economic development and population growth.[9,10]

Six benchmark scenarios were developed by the Fourth Assessment – A1F1, A1T, A1B, A2, B1 and B2 – which were taken to be representative of the wide range of possible world development pathways and resultant emissions in the next century and to encompass a significant proportion of the underlying uncertainties in the main driving forces of emissions (see Figures 3 and 4).[10] Scenarios A1F1 and A1B form the "worst-case" scenarios with high greenhouse gas emissions over the next century; scenarios B1 and B2 simulate reducing emissions by 2100, with lower population growth and the

---

A1 scenario family: a future world of very rapid economic growth, global population that peaks in mid-century and declines thereafter, and rapid introduction of new and more efficient technologies.

A2 scenario family: a very heterogeneous world with continuously increasing global population and regionally oriented economic growth that is more fragmented and slower than in other storylines.

B1 scenario family: a convergent world with the same global population as in the A1 storyline but with rapid changes in economic structures toward a service and information economy, with reductions in material intensity and the introduction of clean and resource-efficient technologies.

B2 scenario family: a world in which the emphasis is on local solutions to economic, social, and environmental sustainability, with continuously increasing population (lower than A2) and intermediate economic development.

Source: Nakicenovic et al. 2000 (ref. 10).

Figure 3   Outline of the main IPCC scenario storylines.

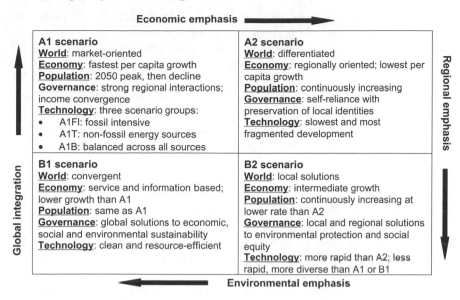

Source: Nakicenovic et al. 2000 (ref. 10).

Figure 4   Key differences between the potential "pathways" of global socio-economic
development used in the IPCC AR4 emissions scenarios.

development of lower-carbon energy sources. Which scenario is most realistic
for future emissions depends critically upon whether economic development
remains the dominant global aspiration in the future and whether present
socio-economic globalisation will intensify or wane (see Figure 4).[9,10]

Likely climate change from each of the scenarios is predicted by forcing an
ensemble of Global Circulation Models (GCMs) with predicted concentrations
of long-lived greenhouse gases (LLGHGs) from the emissions scenarios.[4,11,12]
This approach enables differences between the climate projections modelled for
each emissions scenario to be quantified and enables clear identification of
regions in which there is large uncertainty in projected climate change.

2.2   Key Uncertainties in Climate Projections

2.2.1   General Uncertainties in Climate Projections
Although climate projections are used to drive many adaptation and development discussions, there are significant limitations to the predictive capability
of GCMs in assessing likely climate change in many regions worldwide, particularly in data-poor regions such as Africa.[4] Uncertainties in climate modelling are, at least in part, due to the exclusion of several key, but poorly
understood feedback processes between global climate and the carbon cycle in
current GCMs. Feedback between the climate and carbon cycle is likely to

influence significantly how much climate change will result from different levels of emissions. However, our poor understanding of some of these highly complex processes driving land and ocean uptake of carbon means key feedback processes are excluded from most current GCMs. Consequently, many GCMs model future climate change and the global carbon cycle as completely uncoupled systems, so that even feedback effects between vegetation and land-use are excluded.[4] The new process-based C^4MIP carbon-cycle GCM suite is one of the first attempts to simulate some level of coupling between CO_2 loading and the global climate cycle, and the results have so far indicated that previous estimates of emissions cuts required to achieve a stabilisation of atmospheric CO_2 concentrations in the Earth's atmosphere by 2100 have been conservative.[13,14]

Significant uncertainty in climate change predictions also arises from the underlying uncertainties in the emissions scenarios used to force the climate models and in the assumptions made to simulate physical processes of the climatic system within GCMs. The exclusion of daily or inter-annual climatic variability within GCMs is particularly important, as it is this very short-term climatic variability which is thought to be highly important in simulating the effect of intense rainfall events and the future frequency of droughts.

2.2.2 Uncertainties in African Climate Projections

The margin of error in climate predictions for Africa is large: up to 90% of current GCMs cannot replicate accurately past or present climatic conditions observed within large parts of Africa – particularly in sub-Saharan Africa. Such large errors in the climate models suggest important climatic processes are not being modelling within the GCMs.[4] Whilst current temperature changes can be replicated, rainfall can be overestimated by up to 20% in sub-Saharan Africa and sea-surface temperatures (which have a significant effect on rainfall patterns) can be overestimated by 1–3 °C by current GCMs.[4]

Modelling climate systems within Africa is particularly difficult due to the lack of observed data from the continent and the complexity of the continent's climate. Rainfall patterns are dominated by seasonal migration of the tropical rain belts. Small shifts in the position or the timing of the movements of these rain belts will result in large local changes in rainfall. Being able to simulate such processes within climate models is essential for the accurate predictions of the effect of global climate change in Africa. However, as yet, less than half of the climate projections agree on what rainfall change can be expected with seasonal movement of the rain belts. The uncertainty is greatest within the western Sahel (10–18 °N, and 17.5–20 °E) with some GCMs predicting significant drying whilst others simulate a progressive wetting with an expansion of vegetation into the Sahara.[8] New GCM models are being developed, but it is unlikely that climate models will have the capacity to model accurately the required level of climatic complexity for reliable climate projections in the near future.

2.3 Projected Climate Change in Africa

Despite the uncertainties, climate change, both globally and within Africa, is likely to represent an intensification of present climatic variability rather than a catastrophic change in the mean climate state. Best-estimate projections of the IPCC Fourth Assessment Report indicate that the global mean temperature will rise between 1.8 °C under scenario B1 and up to 4 °C under scenario A1F1 by 2100, compared with the 1980–1999 period.[4,15] Warming over land will be greater than the global mean temperature rise, due to less water availability for evaporative cooling and the smaller thermal inertia of the atmosphere as compared to the oceans.[4] As a result of this effect and changes to sea surface temperatures, atmospheric circulation and land-use patterns, the near-surface temperature rise in Africa for the same period is projected to be 3–4 °C (roughly 1.5 times the global mean response; see Figure 5).[8]

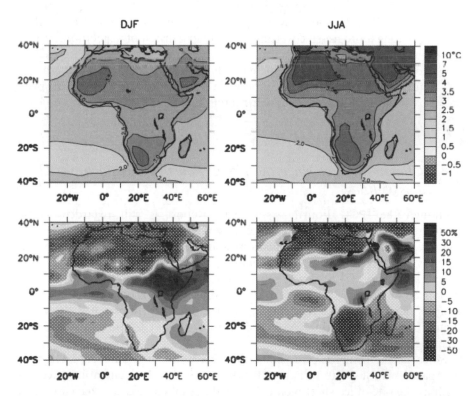

Figure 5 Most likely temperature and precipitation changes modelled within Africa within winter (left) and summer (right) months. DJF refers to months December, January, February, whilst JJA refers to the months June, July, August. Temperature changes are show in the upper panel in degrees Celsius, whilst the predicted percentage change to rainfall is shown in the lower panels.

A warmer global climate will result in many other climatic changes:

- Warmer atmospheric conditions will lead to an intensification of the hydrological system, so that rainfall events will be of greater intensity and of greater spatial and temporal variability.[8] Extreme events, such as tropical storms, floods and droughts are projected to increase in both frequency and intensity.
- Seasonal rainfall patterns in Africa, which are presently dominated by the seasonal migration of the tropical rain belts, will be exaggerated, with small shifts in the position of these rain belts resulting in large local changes in rainfall.[9] However, there is considerable uncertainty within the GCMs as to how rainfall is likely to change in the future.
- Overall, GCMs generally predict rainfall to decrease by up to 20% in northern Africa, to decrease by up to 30% in southern Africa, and to increase by *ca.* 7% within central and eastern Africa (see Figure 5).[4,8] Uncertainty is greatest for the western Sahel (10–18 °N, and 17.5–20 °E), with some GCMs predicting significant drying and others simulating a progressive wetting, with an expansion of vegetation into the Sahara.[8]
- More than half of the predicted reduction in rainfall in northern and southern Africa is modelled to occur in the northern hemisphere in spring as a result of a delay in summer rainfall.[8]

2.4   Climate Science since the IPCC Fourth Assessment: 4 °C Possibilities

Although the IPCC Fourth Assessment Report is still regarded as the authoritative work on climate change, observed data since 2000 indicates that the IPCC scenarios of future emissions and "likely" climate change are conservative.[13,16] The Fourth Assessment Report concluded that a mean global temperature rise of 1.8 °C was most likely, but population growth, intensification of economic globalisation and a failure to decarbonise global energy sources has meant fossil-fuel emissions have consistently matched those predicted by the "worst-case" (A1F1) scenario of the IPCCs Fourth Assessment (see Figure 6).[7,17,18]

Since 2000, fossil fuel emissions have increased at a rate of 3.4% yr^{-1}, compared with just 1.0% yr^{-1} in the 1990s (ref. 19). Key climatic indicators, such as the rate of ice-sheet melt and global ocean temperature, have exceeded those predicted to be likely and sea level is now expected to rise by more than 1 m by 2100 due to thermal expansion alone (0.5 m more than that predicted by the IPCC in 2007), and by up to a further metre due to current rates of melting of the Arctic and Greenland ice sheets.[16] There is also evidence to suggest that the land and ocean sinks of atmospheric CO_2 are becoming weakened and not keeping up with current increases in CO_2 emissions.[14,19] A weakening of CO_2

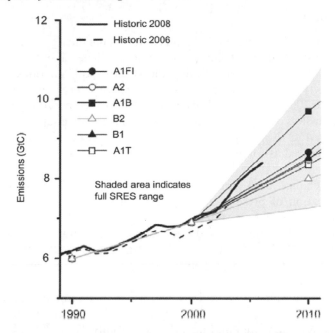

Figure 6 Present trajectory of emissions (as Gigatons Carbon) since 2000 has consistently matched the worst-case emissions scenario of the IPCC Fourth Assessment Report. The shaded area indicates the range of emissions predicted by the IPCC in the Special Report on Emissions Scenarios (SRES) in the 4th Assessment Report. Source: Van Vuuren and Riahi 2008 (ref. 7)

sinks could lead to much greater warming from emission increases than observed at present.[19]

Based on observational evidence since 2000, there is an emerging consensus that a global mean temperature rise of 2.4 °C is likely, regardless of any emissions cuts, and that a global mean temperature rise of 4 °C is more likely than not by 2100 (ref. 6,7,16,19). Recent modelling work using the HadCAM GCM suite suggests a 4 °C rise in global mean temperature would result in a rise in near-surface temperatures in Africa of between 5.0 to 7.5 °C by 2100 (ref. 20,21). Such a strong warming of the atmosphere would have a significant impact on processes in the hydrological system in Africa[4] and it is increasingly thought that we need to prepare for greater climatic uncertainty, based on a 4 °C rise in global mean temperature. In response, the Fifth Assessment of climate change, due to be published by the IPCC in 2013, is to be based on a revised set of emissions scenarios which will include a wider range of possible future emissions.[22] The scenarios will also be generated using a more cause-and-effect approach, to enable better simulation of the feedbacks between increased emissions, potential climate change and adaptation.[22]

2.5   Summary

The IPCC Fourth Assessment of climate change is the most valid assessment to date, but the predictions of climate change are increasingly thought to be conservative. Despite the significant uncertainty still surrounding climate change projections, there is an emerging consensus based on observational evidence since 2000 that there will be a global mean temperature rise of 2.4 °C by 2100, regardless of any future emissions cuts, and that adaption should now be based on a 4 °C rise in global mean temperate. Key facets of climate change can be summarised as follows:

- Climate change, both globally and within Africa, is likely to represent an intensification of present climatic variability rather than a catastrophic change in mean climate state.
- It is highly likely that by 2100 near-surface temperatures will be 3–4 °C higher over Africa (possible 5–7.5 °C) and, even under the most conservative climate change prediction, global sea-level rise is likely to be more than 1 metre.
- Warmer atmospheric conditions will lead to an intensification of the hydrological system, so that rainfall events will be of greater intensity and of much greater spatial and temporal variability.
- Generally, rainfall is predicted to decrease by up to 20% in northern Africa, to decrease by up to 30% in southern Africa and to increase by *ca.* 7% within central and eastern Africa (see Figure 5). However, up to 90% of GCMs cannot accurately replicate current climatic conditions in Africa, so there is significant uncertainty in rainfall projections, particularly within the western Sahel where it is still unclear whether rainfall will increase or decrease.
- Climate modelling uncertainties are large in Africa. Up to 90% of GCMs cannot accurately replicate past or present climatic conditions within sub-Saharan Africa, suggesting key climatic processes are being omitted from models.

3   Impacts of Climate Change on Rural Water Supply in Africa

3.1   A Framework for Discussion

Predicting the effects of climate change on rural water supplies in Africa is difficult, not least because of the considerable uncertainty in climate change predictions and the even greater uncertainty within derived hydrological models.[23] There are also large data and knowledge gaps in existing run-off and recharge processes within Africa.[24] To help deal with this level of uncertainty in climate science, it is useful to adopt a three-sided approach to examine how climate change is likely to affect rural water supplies, focusing on water availability, water use and the ability to access available water.[5] Ultimately it is the interplay between these three factors which will determine future water security in Africa.[3,5]

3.2. Likely Impact of Climate Change on Available Water Resources

3.2.1 General

Within a warmer climate, there will be higher evaporative demand, higher sea and land surface temperatures and consequently existing climatic variability in Africa will be intensified, so that rainfall will occur in more intense events of higher spatial and temporal variability and dry periods will be both more prolonged and more frequent.[4,25] Whilst this climate change is highly likely, how it will translate to changes in effective rainfall and the partitioning of this effective rainfall between different water resources through altered patterns of surface run-off, soil moisture and groundwater recharge, is unclear.[26] Collectively, there are too many uncertainties, and the processes and feedback processes involved too complex for the effect of climate change on hydrological systems to be modelled adequately at present.[26] It is, therefore, very difficult to predict the likely impacts of climate change on rural water supplies from climate model projections. However, examining how the different water resources in Africa respond to existing climatic variability can provide an insight into the likely effects of climate change. Improved supplies generally rely on groundwater resources, whilst for unimproved supplies surface water and very shallow, perched groundwater (< 10 m deep) resources are important.

3.2.2 Surface Water Resources

Surface water resources in Africa are already strongly seasonal and variability in river flow in most regions is marked as a result of present climatic variability (see Figures 7 and 8).[27–29] Superimposed on these seasonal variations are "natural" variations, in which drought and flood are already part of the existing hydrological system.[31] Increased intensity and irregularity of rainfall with future climate change will mean inter-annual variability of river flows is likely to increase, such that rivers will be increasingly seasonal and flashy.

Figure 7 Strong variability of surface water availability already exists in sub-Saharan Africa. The difference in seasonal flow of River Oju, SE Nigeria: mid-wet season (left), mid dry season 4 months later (right).

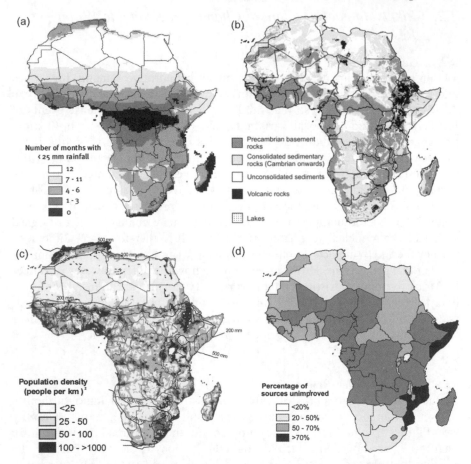

Figure 8   Climatic and hydrogeological maps of Africa. Map (a) illustrates the average
length of the dry season in Africa (1961–1996), calculated using data from
New *et al.* 1999 (ref. 30); map (b) indicates the main hydrogeological
environments of Africa;[3] map (c) displays the main rainfall-recharge zones
relative to the population density in Africa;[3] and map (d) shows the percentage of unimproved supplies within rural Africa over the continent using
data from the JMP 2008 (ref. 1).

Flood events within the wet season may become more common and an
increased proportion of "available" surface water will be lost or redistributed in
peak river discharges, reducing year-round access.

In the long term, change to surface water availability will be entirely
dependent on how changes in rainfall patterns and increased evaporative
demand translate to shifts in soil moisture deficits and surface water run-off.[8,26]
At present, quantifying such changes is beyond present prediction capabilities,
but it is highly likely that surface water resources will become increasingly
unreliable.[8,32]

3.2.3   Groundwater Resources

Most improved water supplies in rural Africa depend on groundwater.[3] As rainfall and surface waters become less reliable, the demand on groundwater-based supplies is likely to increase further.[3] Unlike surface water, groundwater is less responsive to short-term climatic variability and will be buffered to the effects of climate change in the near-term as a result of the storage capacity of the aquifer. The potential long-term impact of climate change on the availability of groundwater is, however, largely unknown, not least because of the complexity of recharge processes in Africa, which are poorly constrained at present, even without the complications of climate change.[33,34]

Recharge

Climate change is likely to modify groundwater recharge patterns, as changes in precipitation and evaporation translate directly to shifts in soil moisture deficits and surface water run-off.[3,35] Increases in rainfall intensity and evaporative demand will, more likely than not, result in increased irregularity of groundwater recharge.[26] Recent modelling work within humid regions of Africa, has stressed the link between rainfall intensity and recharge, and it is possible that greater rainfall intensity predicted with climate change might in fact lead to increases in recharge by up to 50–200%.[36–38] There is no simple, direct relationship between rainfall and recharge, and recharge patterns will additionally be affected by soil degradation and vegetation changes that are likely with increased climatic variability. Even though rainfall intensity is likely to increase in the future, soil degradation and vegetation changes might in fact mean more rainfall becomes surface run-off, so that recharge will decrease.[36,39–41] There is still, therefore, a large degree of uncertainty as to what the final effect of climate change will be on recharge patterns, particularly within semi-arid regions, as a result of the complexity of the processes.

Aquifer Storage

Recent studies of rural water use and the response of groundwater supplies have indicated that, even with reduced and more irregular recharge patterns, groundwater resources at depth (probably deeper than 10–20 m) in many aquifers will generally be of sufficient storage capacity to remain a secure water resource for the domestic water need in Africa. Groundwater-based domestic supplies require little recharge (< 10 mm), as rural domestic water use is low (average hand pump yield is $5–10 \, m^3 \, d^{-1}$).[3,42,43] Based on these recharge values and some preliminary modelling work, MacDonald *et al.* (2009; ref. 3) suggest that it is low-storage aquifers in areas where annual rainfall is currently between 200 and 500 mm that are most at risk (see Figure 8). Although recharge processes are highly complex, the limited available data suggest that recharge becomes much less reliable at rainfall less than 500 mm per annum.[44–46]

However, it is only *improved* water supplies which access deeper, more sustainable groundwater resources. Instead, the majority of the rural population relies on unimproved supplies which use shallow groundwater sources

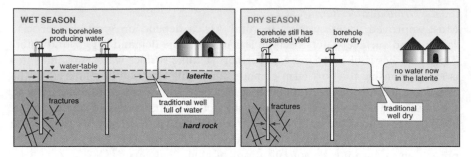

Figure 9   The shallow, perched groundwater resource held within laterite soil often fails in the dry season, leading to the failure of unimproved water supplies.[53] In contrast, groundwater resources at depth (more than 20 m below ground surface) are modelled to be of sufficient storage capacity to support rural domestic water demand, based on a hand pump with a yield of $5 \, m^3 \, d^{-1}$.

(generally less than 10 m deep) within deep soils (see Figure 8). This groundwater is often ephemeral and the soil layers are of much lower storage capacity than aquifers at depth; the shallow groundwater relies on frequent regular recharge for the resource to be sustainable.[3] During extended dry seasons under existing climatic variability, the shallow groundwater resource already often fails (see Figure 9).[47] Increased rainfall variability predicted with climate change is likely to mean unimproved water supplies in low storage regolith (weathered soil material) aquifers could fail more often more of the time.

Quality of Water Resources

Even without climate change, there are significant threats to the quality of surface and shallow groundwater sources as a result of the lack of sanitation in rural Africa (less than 70% of the population has access to sanitation), the use of on-site sanitation and the growing use of fertilisers to increase crop yields.[1,48] Intense rainfall events can flush contamination from soil into rivers and groundwater. Once within groundwater, pathogenic contaminants can be transported significant distances (up to 1 km) through the permeable regoliths whilst still virulent.[48–50] Within the wet season, high groundwater levels (often less than 10 m below ground surface) also mean pathogens (and other suspended contaminants) can enter shallow groundwater directly from the base of latrines and other conduits, making the shallow groundwater source highly vulnerable to contamination.[3]

Climate change may exacerbate these existing water quality issues, so that the quality of surface water and shallow groundwater may be further impacted.[9,26] Increased flooding of latrines and unimproved sources could lead to a rise in diarrhoeal disease and infant mortality, and warmer water temperatures could lead to greater transmission of disease, for example.[51] Reduced functioning of water supplies during extended droughts could increase the burden of disease.

In regions where surface water and groundwater recharge are projected to decrease, general inorganic water quality may also decrease due to the lower dilution capacity of the water resources.[26] However, predictions of how, and by how much, future water quantities will change are so uncertain that the potential for "dilution" of the contaminant loading if water quantities were to increase cannot be relied upon to counter the flushing effect of more intense rainfall.[26]

3.3 Access to Reliable Water Supplies

Despite the considerable climatic variability which currently exists across much of Africa, the key determinant of existing water security in rural Africa is reliable *access* to water resources and not actual water resource availability itself.[3,5] Research from Ghana, Nigeria, Malawi, South Africa and Ethiopia over the last 20 years provides some insights.[5,52,53] During droughts, unimproved supplies commonly fail as a result of high demands outstripping the limited capacity of unimproved sources.[43,54] The shallow groundwater resource, which unimproved supplies invariably access, requires regular annual recharge to ensure sustainable supply as a result of the low storage capacity of the regolith. Within dry seasons, the shallow regolith aquifers are often depleted and unimproved supplies fail. Only supplies which access larger, deeper groundwater resources (approximately >20 m) have been seen to be reliable (see Figure 9). Increasing access to these more secure groundwater resources could, therefore, go a significant way to mitigating existing and future water insecurity within rural Africa.

Improved supplies are not infallible, however, and they can fail in droughts when demand on improved supplies is high as a result of widespread failure of unimproved supplies.[5] Prolonged pumping can cause mechanical failure of hand pumps, especially if water levels in boreholes have fallen and the pump lift is increased.[55] Mechanical failure is all the more likely if there has been a lack of pump maintenance.[43] Improved sources can therefore fail, even if sufficient water is regionally available for the high demand. This demand-driven failure of improved supplies is most likely in aquifers of limited permeability and storage and where there are few other improved sources locally to help disperse the demand across several improved wells.[43,54]

Increasing the resilience of rural water supplies to climate change should, therefore, be achieved by adopting actions already required to improve water security in Africa.[3,33] Increasing access to secure water supplies is a clear priority. But what technologies will be most reliable? Which will cope best with a changing climate? Table 1 outlines the main technology choices available in rural Africa. Technologies that rely on small streams and rivers, or ephemeral shallow groundwater resources, are likely to be the least reliable and are of greatest vulnerability to contamination. Those which abstract water from large surface water sources or deeper groundwater are likely to be best able to cope with climate change.[55]

Table 1 The main water-supply technology choices for rural Africa. The effects of predicted climate change in Africa for each technology are outlined, alongside possible mitigation measures.

Technology	Description	Climate risks	Possible impacts	Responses
Rainwater harvesting.	Collecting water from rainwater and storing in tanks – can be household or community.	There may be fewer rainy days – and longer drought periods. Rainfall events may be more intense.	Larger storage may be required to provide storage for the longer dry days.	Build in redundancy for potential reduced rainfall and longer dry seasons.
			Danger of damage and contamination from flooding.	Ensure protection against flooding.
Reticulated schemes from small rivers and dams.	Pumped schemes to villages and small towns based on small dams or river abstraction.	Changed seasonality of runoff, peak flows and sediment load.	Lower and less certain flows. Possible increased sedimentation.	Design to a higher capacity.
			Dams may be filled with sediment – possibility of failure.	Build in mechanisms for dealing with increased sedimentation.
				Conjunctive use of surface and groundwater to increase adaptability to change.
Shallow family wells.	Wells less than 10 m deep – dug by hand and often unlined.	More intense rainfall, longer dry season.	Increased contamination of sources.	Should generally not be promoted as improved water supplies.
			More likely that sources will fail.	

Improved hand-dug wells.	Hand-dug wells, often >10 m deep, lined with concrete and capped at the surface.	More intense rainfall, longer dry season.	Increased risk of contamination. More likely that sources will fail.	Hand-dug wells should be tested at the peak of a normal dry season. They should be sited in productive parts of the aquifer and deep enough to intersect groundwater below 10 m. There should be an emphasis on casing out shallow layers and runoff.
Protected spring supplies.	Perennial springs where the source is protected and piped to a standpipe.	Longer dry season – more intense rainfall.	Possibility of contamination – particularly if in more urban setting. Springs may be less reliable in longer dry seasons.	More thorough investigation of seasonal spring flow and contamination pressures in catchment. Build in greater redundancy.
Boreholes.	Boreholes, 20–60 m deep, with hand-pump mechanism to abstract water.	Longer dry season – more intense rainfall.	Higher demand within extended dry seasons may cause source failure and, in some cases, depletion of water resource. High demand can lead to mechanical failure.	To improve reliability of water supply, ensure boreholes are sited in most productive part of aquifer and are of higher storage capacity. It is also important to improve maintenance of the hand-pumps – particularly within the dry season.

Table 1 Continued.

Technology	Description	Climate risks	Possible impacts	Responses
			Risk of supply contamination from very shallow layers of source during intense rainfall.	Ensure shallow layers of groundwater source are cased out to prevent contamination of the supply.
Large piped schemes from large dams and rivers.	Capital-intensive schemes to large towns and cities.	Increased demand in cities. Changes in runoff and sedimentation.	Larger storage should be able to cope with climate fluctuations.	Although this water resource may be more resilient to climate change, effects of demand and possibilities of reliability may make it vulnerable.
			Large increase in demand may lead to failure.	Consideration should be given to conjunctive use, backup and designing to cope with higher demand.

In summary, increasing access to the most appropriate water resources could mitigate the effects of increased climatic variability; climate change need not necessarily lead to catastrophic, continent-wide failure of rural water supplies, as is too commonly portrayed in the media.[5] Several actions would help increase water security:

- Increasing water supply coverage to meet the Millennium Development Goals for water, reducing dependence on shallow unimproved sources.
- Targeting large reliable water resources, such as deeper groundwater.
- Matching the water supply technology to groundwater conditions and siting sources in the most productive parts of the aquifer.
- Maintaining water sources so that more are operational at the outset of drought periods.
- Ensuring water sources are protected from periodic flooding and contamination.

3.4 Changing Water Demands

Whilst climate change is significant and will undoubtedly affect the availability of some water resources in rural Africa, it is not happening in isolation and other pressures are likely to have a greater impact on water security over the next 50 years – most notably, population growth, land-use change and agriculture demand.[56] Unlike climate change, the prospects of demographic change in Africa in the 21st Century are known with some certainty.[56] Africa has the highest population growth rate in the world at present (1.6–3.1% per year) and the population is set to increase from 900 million in 2009 to approximately 2 billion by 2100 (ref. 57,58). The increase in domestic and agricultural demand for freshwater which will follow this population growth are likely to outstrip any effects of climate change on water security in Africa[3,56,59]

In rural Africa, where the population is dispersed, the increase in domestic water demand predicted with population growth is unlikely to pose a significant threat to the sustainability of renewable water resources, as long as the rural population remains dispersed and water demand limited to 20 litres per person per day.[56] It is estimated population growth in rural areas by the end of the 21st Century will lead to an increase in water demand equivalent to less than 0.5 mm water a year across the continent; less than 1% of total annual recharge estimated to presently occur over Africa.[56] However, increased water demand in the agricultural sector to meet rising food demands with population growth will place a much greater demand on available water resources.[33,56] Over-exploitation and widespread depletion of the groundwater resource could easily occur in the absence of careful management.

Future food security with predicted population growth can only be met with increased use of irrigation and careful water management.[56] Already there are concerns about whether food production from rain-fed and irrigated systems can be increased to meet the growing food demand, and increased land degradation and more prolonged dry periods with climate change will only exacerbate current difficulties.[3] Indeed, it is estimated that climate change could cause rain-fed cereal yields to decrease by up to 50% (ref. 8).

Increasing the level of irrigation within Africa could help increase food production, just as groundwater-fed irrigation has proved a viable way of increasing agricultural output to support rising demands in Asia. However, this development of groundwater was only possible in Asia because of the existence of extensive, productive aquifers (*e.g.* the deep sedimentary aquifer in Gujurat and the basaltic aquifers in Deccan), and the same development could not be supported by the lower-yielding aquifers of limited storage capacity which underlie 80% of Africa's land area.[35,55] Whilst the lower yielding aquifers in Africa are thought to represent a secure source of water for dispersed domestic water demands, the aquifers are of insufficient storage and permeability to be able to support very high domestic and agricultural water demands.

Consequently, intensive groundwater-based irrigation is unlikely to be an appropriate Africa-wide strategy, and even small-holder irrigation schemes in rural areas need to be managed carefully so as to ensure sustainable exploitation of the groundwater source.[55] Increasing groundwater abstraction ten-fold to meet agricultural demand near urban areas could threaten the sustainability of domestic water supplies and could lead to widespread over-exploitation – particularly in the more-productive aquifers where regional drawdown of ground-water-levels are more likely.[3] Indeed, in the higher-yielding sandstone and unconsolidated sedimentary aquifers in areas of north Africa, where ground-water development supports roughly 30% of agricultural production, agriculture is already mining non-renewable reserves and nearing the limit of expansion.[60]

4   Summary

Significant uncertainty still surrounds present climate projections. However, there is an emerging consensus, based on observational evidence since 2000, that global mean temperature will rise by 2.4 °C by 2100, regardless of any emission cuts, and that adaptation should now be based on a 4 °C rise in global mean temperature.[7,16,19] One of the key uncertainties surrounding the impacts of a changing climate in Africa is the effect that it will have on the sustainability of rural water supplies. Of Africa's population of 900 million, roughly 60% live in rural areas and most (perhaps 80%) rely on groundwater-based community or household supplies for domestic and other water needs. Understanding the impacts of climate change on groundwater resources is, therefore, of critical importance, yet is often ignored in development debates. Below are some of the key issues that need to be considered:

1. Those relying on unimproved water sources (300 million in rural Africa) are likely to be the most impacted by climate change. This is because unimproved sources often tap ephemeral water resources such as very shallow groundwater, ponds and small streams.
2. Improved rural water supplies in Africa are overwhelmingly dependent on groundwater and dependence is likely to increase as surface water

resources become more unreliable. A key advantage of groundwater-based supply is reliability, both for domestic and productive uses.
3. Climate modelling uncertainties, combined with rapid socio-economic change, make predicting water resources in the future difficult. While there is good confidence in temperature projections, rainfall scenarios remain uncertain, as do impacts on groundwater recharge. Demand-side pressures on water resources are more clear cut and may dwarf the impacts of climate change.
4. Is there a future problem? Yes, but it is important to emphasise that climate change will not lead to the continent-wide failure of improved rural water sources. This is because groundwater-based domestic demand requires very little recharge. However, a significant minority of people could be affected if rainfall declines in those areas with limited groundwater storage, especially if the frequency of droughts increases.
5. In most areas, the key determinant of water security will continue to be access rather than availability. Extending access, and ensuring that targeting and technology decisions are informed by an understanding of groundwater conditions, is becoming increasingly important.
6. Accelerating groundwater development for irrigation could increase food production, raise rural incomes and reduce vulnerability. However, *ad hoc* development could threaten domestic supplies and, in some areas, lead to groundwater depletion.

References

1. *JMP Global Water Supply and Sanitation 2008 Report,* Joint Monitoring Programme WHO/UNICEF, World Health Organisation, Geneva, 2008.
2. A. M. MacDonald, J. Davies, R. C. Calow and J. Chilton, *Developing Groundwater A Guide for Rural Water Supply*, ITDG Publishing, Rugby, UK, 2005, pp. 358.
3. A. M. MacDonald, R. C. Calow, D. M. J. MacDonald, G. W. Darling and B. É. Ó Dochartaigh, *Hydrol. Sci. J.*, 2009, **54**, 690–703.
4. G. A. Meehl, T. F. Stocker, W. D. Collins, P. Friedlingstein, A. T. Gaye, J. M. Gregory, A. Kitoh, R. Knutti, J. M. Murphy, A. Noda, S. C. B. Raper, I. G. Watterson, A. J. Weaver and Z.-C. Zhao, *Global Climate Projections*, in *Climate Change 2007: The Physical Science Basis. Contribution of Working Group I to the Fourth Assessment Report of the Intergovernmental Panel on Climate Change*, ed. S. Solomon, D. Qin, M. Manning, Z. Chen, M. Marquis, K. B. Averyt, M. Tignor and H. L. Miller, Cambridge University Press, Cambridge, UK and New York, NY, USA, 2007.
5. R. C. Calow, A. M. MacDonald, A. L. Nicol and N. S. Robins, *Ground Water*, 2010, **48**, 246–256.
6. V. Ramanthan and Y. Feng, *Proc. Natl. Acad. Sci. U.S.A.*, 2008, **105**(38), 14245–14250.
7. D. P. Van Vuuren and K. Riahi, *Climate change*, 2008, **91**, 237–248.

8. M. Boko, I. Niang, A. Nyong, C. Vogel, A. Githeko, M. Medany, B. Osman-Elasha, R. Tabo and P. Yanda, Africa, in *Climate Change 2007: Impacts, Adaptation and Vulnerability, Contribution of Working Group II to the Fourth Assessment Report of the Intergovernmental Panel on Climate Change*, ed. M. L. Parry, O. F. Canziani, J. P. Palutikof, P. J. van der Linden and C. E. Hanson, Cambridge University Press, Cambridge UK, 2007, 433–467.

9. B. Bates, Z. Kundzewicz, S. Wu and J. Palutikof, *IPCC: Climate Change and Water, IPCC Working Group II, Technical Paper of the Intergovernmental Panel on Climate Change*, IPCC Secretariat, Geneva, 2007, pp. 210.

10. N. Nakicenovic, J. Alcamo, G. Davis, B. de Vries, J. Fenhann, S. Gaffin, K. Gregory, A. Grübler, T. Yong Jung, T. Kram, E. L. La Rovere, L. Michaelis, S. Mori, T. Morita, W. Pepper, H. Pitcher, L. Price, K. Riahi, A. Roehrl, H.-H. Rogner, A. Sankovski, M. Schlesinger, P. Shukla, S. Smith, R. Swart, S. van Rooijen, N. Victor and Z. Dadi, IPCC Special Report on emissions scenarios, A *Special Report of IPCC Working Group III*, 2000.

11. N. W. Arnell, *Global Environ. Change*, 2004, **14**(1), 31–52.

12. D. A. Randall, S. Wood, R. Bony, T. Colman, J. Fichefet, V. Fyfe, A. Kattsov, J. Shukla, J. Srinivasan, R. J. Stouffer, A. Sumi and K. E. Taylor, Climate Models and Their Evaluation, in *Climate Change 2007: The Physical Science Basis, Contribution of Working Group I to the Fourth Assessment Report of the Intergovernmental Panel on Climate Change*, ed. S. Solomon, D. Qin, M. Manning, Z. Chen, M. Marquis, K. B. Averyt, M.Tignor and H. L. Miller, Cambridge University Press, Cambridge, UK and New York, NY, USA, 2007.

13. M. Meinshausen, N. Meinshausen, W. Hare, S. C. B. Raper, K. Frieler, R. Knutti, D. J. Frame and M. R. Allen, *Nature*, 2009, **458**(30), 1158–1163.

14. P. Friedlingstein, P. Cox, R. Betts, L. Bopp, W. von Bloh, V. Brovkin S. Cadule, M. Doney, M. Eby, I. Fung, J. Bala, J. John, C. Jones, F. Joos, T. Kato, M. Kawamiya, W. Knorr, K. Lindsay, D. Matthews, T. Raddatz, P. Rayner, C. Reick, E. Roeckner, K.-G. Schnitzler, R. Schnur, K. Strassmann, A. J. Weaver, C. Yoshhikawa and N. Zeng, *J. Climate, Am. Meterol. Soc.*, 2006, **19**, 3337–3353.

15. M. L. Parry, O. F. Canziani, P. Palutikof, P. J. van der Linden and C. E. Hanson, *Climate Change 2007: Impacts, Adaptation and Vulnerability, Contribution of Working Group II to the Fourth Assessment Report of the IPCC*, Cambridge University Press, Cambridge, UK, 2007.

16. K. Richardson, W. Steffen, H. J. Schnellnhuber, J. Alcamo, T. Barker, D. M. Kammen, R. Leemans, R. Liverman, M. Munasinge, B. Osman-Elasha, N. Stern and O. Waever, *International Scientific Congress Synthesis Report, Climate Change, Global Risks, Challenges and Decisions, Copenhagen 2009*, Copenhagen University Press, Denmark, 2009.

17. M. R. Raupach, G. Marland, P. Ciais, C. Le Quere, J. G. Canadell, G. Klepper and C. B. Field, *Proc. Natl. Acad. Sci. U.S.A.*, 2007, **104**(24), 10288.

18. R. K. Pachauri and A. Reisinger, Contribution of Working Groups I, II and III to the Fourth Assessment Report of the Intergovernmental Panel on Climate Change, *IPCC Climate Change 2007: Synthesis Report*, 2007.
19. C. Le Quéré, M. R. Raupach, J. G. Canadell, G. Marland, L. Bopp, P. Ciais, T. J. Conway, S. C. Doney, R. A. Feely, P. Foster, P. Friedlingstein, K. Gurney, R. A. Houghton, J. I. House, C. Huntingford, P. E. Levy, M. R. Lomas, J. Majkut, N. Metzl, P. E. Ometto, G. P. Peters, C. I. Prentice, J. T. Randerson, S. W. Running, J. L. Sarmiento, U. Schuster, S. Sitch, T. Takahashi, N. Viovy, G. R. van der Werf and F. I. Woodward, *Nature Geosci.*, 2009, **2**, 831–836.
20. N. W. Arnell, Climate change and water resources a global perspective, in *Avoiding Dangerous Climate Change Proceedings of the Exeter Conference*, ed. H. J Schellnhuber, W. Cramer, N. Nakicenovic, T. Wigley and G. Yohe, Cambridge University Press, Cambridge, UK, 2006, 167–175.
21. D. P. Van Vuuren, M. Meinshausen, G.-K. Plattner, F. Joos, K. M. Strassmann, S. J. Smith, T. M. Wigley, S. C. B. Raper, K. Riahi, F. de la Chesnaye, M. G. J. den Elzen, J. Fujino, K. Jiang, N. Nakicenovic, S. Paltsev and J. M. Reilly, *Proc. Natl. Acad. Sci.U.S.A.*, 2009, **105**(40), 15258–15262.
22. R. Moss, M. Babiker, S. Brinkman, E. Calvo, T. Carter, J. Edmonds, T. Elgizouli, S. Emori, L. Erda, K. Hibbard, R. Jones, M. Kainuma, J. Kelleher, J. F. Lamarque, M. Manning, B. Matthews, J. Meehl, L. Meyer, J. Mitchell, N. Nakicenovic, B. O'Neill, R. Pichs, K. Riahi, S. Rose, P. Runci, R. Stouffer, D. van Vuuren, J. Weyant, T. Wilbanks, J. P. van Ypersele and M. Zurek, Towards New Scenarios for Analysis of Emissions, Climate Change, Impacts, and Response Strategies, in *Technical Summary. Intergovernmental Panel on Climate Change*, Geneva, Switzerland, 2008, pp. 25.
23. J. H. Christensen, B. Hewitson, A. Busuioc, A. Chen, X. Gao, I. Held, R. Jones, R. K. Kolli, W.-T. Kwon, R. Laprise, V. Magaña Rueda, L. Mearns, C. G. Menéndez, J. Räisänen, A. Rinke, A. Sarr and P. Whetton, Regional Climate Projections, in *Climate Change 2007: The Physical Science Basis. Contribution of Working Group I to the Fourth Assessment Report of the Intergovernmental Panel on Climate Change*, ed. S. Solomon, D. Qin, M. Manning, Z. Chen, M. Marquis, K. B. Averyt, M. Tignor and H. L. Miller, Cambridge University Press, Cambridge UK and New York, NY, USA, 2007.
24. S. M. A. Adelana and A. M. MacDonald, in *Applied Groundwater Studies in Africa, IAH Selected Papers on Hydrogeology, 13,* CRC Press, Balkema, Amsterdam, The Netherlands, 2008, 1–7.
25. N. W. Arnell, *Hydrol. Earth Syst. Sci.*, 2003, **7**(5), 619–641.
26. Z. W. Kundzewicz, L. J. Mata, N. W. Arnell, P. Döll, P. Kabat, B. Jiménez, K. A. Miller, T. Oki, Z. Sen and I. A. Shiklomanov, Freshwater resources and their management, in *Climate Change 2007: Impacts, Adaptation and Vulnerability Contribution of Working Group II to the Fourth Assessment Report of the Intergovernmental Panel on Climate Change*, ed. M. L. Parry, O. F. Canziani, J. P. Palutikof, P. J. van der

Linden and C. E. Hanson, Cambridge University Press, Cambridge, UK, 2007, 173–210.
27. K. Tilahun, *Water SA*, 2006, **32**(3), 429–435.
28. D. Conway, *Global Environ, Change*, 2005, **15**(2), 99–114.
29. M. Hulme, *Global Environ. Change*, 2001, **11**, 19–29.
30. M. New, M. Hulme and P. D. Jones, *J. Climate*, 1999, **12**, 829–856.
31. D. Verschuren, K. R. Laird and B. F. Cumming, *Nature*, 2000, **403**, 410–414.
32. Z. W. Kundzewicz and P. Döll, *Hydrol. Sci. J.*, 2009, **54**, 665–675.
33. R. G. Taylor, A. D. Koussis and C. Tindimugaya, *Hydrol. Sci. J.*, 2009, **54**, 655–664.
34. P. Döll and K. Fiedler, *Hydrol. Earth Syst. Sci.*, 2008, **12**, 863–885.
35. S. S. D. Foster, A. Tuinhof and H. Garduño, in *Applied Groundwater Research in Africa*, ed. S. M. A. Adelana and A. M. MacDonald, IAH Selected Papers in Hydrogeology 13, Taylor & Francis, Amsterdam, The Netherlands, 2008, 9–21.
36. L. Mileham, R. G. Taylor, M. Todd, C. Tindimugaya and J. Thompson, *Hydrol. Sci. J.*, 2009, **54**, 727–738.
37. L. Mileham, R. G. Taylor, J. Thompson, M. Todd and C. Tindimugaya, *J. Hydrol.*, 2008, **359**, 46–58.
38. P. M. Nyenje and O. Batelaan, *Hydrol. Sci. J.*, 2009, **54**, 713–726.
39. S. Solomon, D. Qin, M. Manning, M. Marquis, K. Averyt, M. B. M. Tignor, H. Leroy Miller and Z. Chen, *Climate Change 2007: The Physical Science Basis, Contribution of Working Group I to the Fourth Assessment Report of the IPCC*, Cambridge University Press, Cambridge, UK, 2007.
40. B. R. Scanlon, K. E. Keese, A. L. Flint, L. E. Flint, C. B. Gaye, W. M. Edmunds and I. Simmers, *Hydrol. Processes*, 2006, **19**, 3285–3298.
41. P. Döll and M. Flöerke, Global-scale estimation of diffuse groundwater recharge, *Frankfurt Hydrology Paper 03*, Institute of Physical Geography, Frankfurt University, Frankfurt am Main, Germany, 2008.
42. R. C. Carter and A. G. Alkali, *Q. J. Eng. Geol.*, 1996, **29**, 341–356.
43. A. M. MacDonald, J. A. Barker and J. Davies, *Hydrogeol. J.*, 2008, **16**, 1065–1075.
44. J. J. De Vries and I. Simmers, *Hydrogeol. J.*, 2002, **10**, 5–17.
45. W. M. Edmunds, in *Applied Groundwater Research in Africa*, ed. S. M. A. Adelana and A. M. MacDonald, IAH Selected Papers in Hydrogeology, Taylor & Francis, Amsterdam, The Netherlands, 2008, **13**, 305–322.
46. V. H. M. Eilers, R. C. Carter and K. R. Rushton, *Geoderma*, 2007, **140**, 119–131.
47. K. Walraevens, I. Vandecasteele, K. Martens, J. Nyssen, J. Moeyersons, T. Gebreyohannes, F. De Smedt, J. Poesen, J. Deckers and M. Van Camp, *Hydrol. Sci. J.*, 2009, **54**, 690–703.
48. M. Pritchard, T. Mkandawire and J. G. O'Neill, *Phys. Chem. Earth,* 2008; doi:10.1016/j.pce.2008.06.036.
49. R. G. Taylor, C. Tindimugaya, J. Barker, D. Macdonald and R. Kulabako, *Ground Water*, 2010, **48**, 284–294.

50. S. M. A. Adelana, T. A. Abiye, D. C. W. Nkhuwa, C. Tindimugaya and M. S. Oga, in *Applied Groundwater Research in Africa*, ed. S. M. A. Adelana and A. M. MacDonald, IAH Selected Papers in Hydrogeology, Taylor & Francis, Amsterdam, The Netherlands, 2008, **13**, 1–7.
51. A. Costello, M. Abbas, A. Allen, S. Ball, S. Bell, R. Bellamy, S. Friel, N. Groce, A. Johnson, M. Kett, M. Lee, C. Levy, M. Maslin, D. McCoy, B. McGuire, H. Montgomery, D. Napier, C. Pagel, J. Patel, J. A. Puppim de Oliveira, N. Redclift, H. Rees, D. Rogger, J. Scott, J. Stephenson, J. Twigg, J. Wolff and C. Patterson, *Lancet*, 2009, **373**, 1693–1733.
52. R. C. Calow, N. S. Robins, A. M. MacDonald, D. M. J. MacDonald, B. R. Gibbs, W. R. G. Orpen, P. Mtembezeka, A. J. Andrews and S. O. Appiah, *Int. J. Water Resources Dev.*, 1997, **13**(2), 241–261.
53. R. C. Calow, A. M. MacDonald, A. Nicol, N. Robins and S. Kebebe, *The Struggle for Water: Drought, Water Security and Rural Livelihoods*, British Geological Survey Commissioned Report, CR/02/226N, 2006.
54. E. P. Wright, in *The Hydrogeology of Crystalline Basement Aquifers in Africa*, ed. E. P. Wright and W. Burgess, Geological Society London Special Publications, 1992, **6**, 1–27.
55. R. C. Calow and A. M. MacDonald, Overseas Development Institute (ODI) Background Note, March 2009.
56. R. C. Carter and A. Parker, *Hydrol. Sci. J.*, 2009, **54**, 676–689.
57. UN (United Nations) Population Division of the Department of Economic and Social Affairs of the United Nations Secretariat, *World Population Prospects: the 2006 Revision and World Urbanisation Prospects: the 2005 Revision*, 2007.
58. R. C. Carter and J. E. Bevan, in *Applied Groundwater Research in Africa*, ed. S. M. A. Adelana and A. M. MacDonald, IAH Selected Papers in Hydrogeology, Taylor & Francis, Amsterdam, The Netherlands, 2008, 13, 25–42.
59. C. J. Vörösmarty, P. Green, J. Salisbury and R. B. Lammers, *Science*, 2000, **289**, 284–288.
60. W. Gossel, A. M. Sefelnasr, P. Wycisk and A. M. Ebraheem, in *Applied Groundwater Research in Africa*, ed. S. M. A. Adelana and A. M. MacDonald, IAH Selected Papers in Hydrogeology, Taylor & Francis, Amsterdam, The Netherlands, 2008, **13**, 43–65.

The European Water Framework Directive – Chemical Monitoring Programmes, Analytical Challenges and Results from an Irish Case Study

ULRICH BORCHERS,* DAVID SCHWESIG, CIARAN O'DONNELL AND COLMAN CONCANNON

ABSTRACT

The European Water Framework Directive (WFD) is one of the most important legislative instruments in the water field. The overarching objective of the policy is the achievement of a "good status" in all waters of European Member States by the end of 2015. Important milestones include the analysis of pressures and impacts, a characterisation of water bodies and monitoring programmes.

The effectiveness of these monitoring programmes and hence of the overall WFD implementation will depend highly on the ability of Member States' laboratories to measure the status of water bodies. The chemical status of water bodies is linked to compliance with EU Environmental Quality Standards defined in a daughter directive. In order to assess the chemical status, a number of priority substances and priority hazardous substances have to be monitored. Requirements on the analytical methods to be used for chemical monitoring are laid down in terms of technical specifications by another daughter directive.

In some cases, the requirements on analytical methods (*e.g.* limit of quantification and measurement uncertainty) for chemical monitoring under WFD pose a real challenge, even for state-of-the-art analytical

---

*Corresponding author.

Issues in Environmental Science and Technology, 31
Sustainable Water
Edited by R.E. Hester and R.M. Harrison
© Royal Society of Chemistry 2011
Published by the Royal Society of Chemistry, www.rsc.org

techniques. This has initiated a number of European activities in the development of guidance documents, harmonisation efforts and demand-driven research.

The main tasks, challenges and research needs related to chemical monitoring of water bodies under the WFD are presented, including an outlook on how these issues are tackled. This is completed by the presentation of actual chemical monitoring results from surface waters from the Republic of Ireland.

1   The Chemical Monitoring Approach of the WFD

1.1   Basic Principles and Approach

The European Water Framework Directive (WFD)[1] commits European Union member states to achieve a "good status" of all water bodies by 2015. For surface waters, "good status" includes quantitative as well as qualitative aspects, the latter comprising requirements on the ecological as well as the chemical status of water bodies. A strategy for dealing with pollution of water from chemicals is set out in Article 16 of the WFD. As a first step of this strategy, a list of priority substances was adopted (Decision 2455/2001/EC),[2] identifying 33 substances or groups of substances of priority concern in surface waters throughout the European Union due to their widespread use and their high concentrations in rivers, lakes, transitional and coastal waters. This list will have to be reviewed every four years and updated as appropriate. The current list comprises mainly organic compounds, including various pesticides, certain polycyclic aromatic hydrocarbons (PAH), benzene, halogenated solvents, flame retardants, a plasticiser, surfactants and antifouling agents as well as some heavy metals.

In addition, the WFD requires Member States to identify specific pollutants in their river basins and to include them in the monitoring programmes. Monitoring of both WFD priority substances and other pollutants for the purpose of determination of the chemical and ecological status shall be performed according to Article 8 and Annex V of the WFD.

Member States have accentuated the need for more guidance on implementation of monitoring requirements for chemical substances. In line with previous documents under the WFD Common Implementation Strategy (WFD CIS), a guidance document has been developed under the mandate of the European Commission within the Chemical Monitoring Activity (CMA) in the period October 2005 to March 2007 (ref. 3). While not being legally binding, it presents the outcome of the discussion of the CMA working group on how to monitor chemical substances in the aquatic environment. It states best practices, complements existing monitoring guidance and provides links to relevant guidance documents, European and international standards.

The guidance on chemical monitoring of surface waters covers monitoring design relevant to surveillance, operational and investigative monitoring, techniques for sampling and analysis, as well as aspects of analytical quality assurance and control. It represents the current state of technical development in a field that is undergoing permanent changes through ongoing scientific research.

1.2 Environmental Quality Standards (EQS) and Resulting Monitoring Requirements

There is a broad consensus on the existence of potentially serious risks with irreversible consequences from dangerous substances. Hence, environmental quality standards (EQSs) for priority substances and selected other pollutants, and related compliance checking provisions have been established for inland surface waters (rivers and lakes) and other surface waters (transitional, coastal and territorial waters).[4] Two types of EQSs were set:

1. Annual average concentrations (AA-EQS) for protection against long-term and chronic effects; and
2. Maximum allowable concentrations (MAC-EQS) to avoid serious irreversible consequences for ecosystems due to acute exposure in the short term.

In the absence of extensive and reliable information on concentrations of priority substances in biota and sediments at a Community level, and in view of the fact that information on surface water seems to provide a sufficient basis to ensure comprehensive protection and effective pollution control, establishment of EQS values has been, at this stage, limited for the majority of substances to surface water only. However, as regards mercury, hexachlorobenzene and hexachlorobutadiene it is not possible to ensure protection against indirect effects and secondary poisoning by EQSs for surface water alone. Hence, EQSs referring to concentrations in biota have been established for these compounds at Community level. To check compliance with these EQS values, the most appropriate indicator species among fish, molluscs, crustaceans and other biota should be monitored. In order to allow Member States flexibility depending on their monitoring strategy, they should be able either to monitor and apply those EQSs for biota or to establish stricter EQSs for surface water, providing the same level of protection.

Furthermore, Member States should be able to establish EQSs for other specific substances for sediment and/or biota at national level and apply those EQSs instead of the EQSs for water set out in the proposed Directive. Such EQSs should be established through a transparent procedure involving notifications to the Commission and other Member States so as to ensure a level of protection equivalent to that achieved by complying with the EQSs for water. Moreover, as sediment and biota are important matrices for monitoring of those priority substances and other pollutants that tend to accumulate, in order to assess long-term impacts of anthropogenic activity and temporal trends, Member States should ensure that existing levels of contamination in biota and sediments will not increase. In this respect, particular attention should be given to anthracene, fluoranthene, polybrominated diphenyl ethers (PBDEs), C10–C13 chloroalkanes, di(2-ethylhexyl)phthalate, hexachlorobenzene, hexachlorobutadiene, hexachlorocyclohexane, pentachlorobenzene, polycyclic aromatic hydrocarbons (PAHs), tributyltin compounds and metals such as cadmium, lead and mercury.

With the exception of cadmium, lead, mercury and nickel, the EQSs set up in the proposed Directive are expressed as total concentrations in the whole water sample. In the case of metals, the EQSs refer to the dissolved concentration, *i.e.* the concentration in the liquid phase of a water sample obtained by filtration through a 0.45 µm filter.

For metals, the compliance regime is adapted by allowing Member States to take into account natural background levels and bioavailability affected by hardness, pH or other water quality parameters.

Compliance with AA-EQSs requires that, for each representative monitoring point within any given water body, the arithmetic mean of the concentrations measured at different times during the year is below the standard. Compliance with MAC-EQSs means that for each representative monitoring point within any given water body the measured concentration at any time during the year must not exceed the standard.

Under the Water Framework Directive, Member States shall furthermore set quality standards for river basin specific pollutants identified in accordance with that Directive and take actions to meet those quality standards by 2015.

Specific pollutants are defined as substances that can have a harmful effect on biological quality and which may be identified by Member States as being discharged to surface waters in significant quantities. According to the WFD, compliance with EQSs for specific pollutants is part of the ecological status assessment. The WFD provides only an indicative list of such pollutants in Annex VIII including:

1. Organohalogen compounds and substances that may form such compounds in the aquatic environment;
2. Organophosphorous compounds;
3. Organotin compounds;
4. Substances and preparations, or the breakdown products of such, that have been proved to possess carcinogenic or mutagenic properties or properties that may affect steroidogenic, thyroid, reproduction or other endocrine-related functions in or *via* the aquatic environment;
5. Persistent hydrocarbons and persistent and bioaccumulative organic toxic substances;
6. Cyanides,
7. Metals and their compounds;
8. Arsenic and its compounds; and
9. Biocides and plant-protection products.

Generally, candidates for specific pollutants will come from List II of the Dangerous Substances Directive (76/464/EEC),[5,6] which is now integrated in the Water Framework Directive and will be fully repealed in 2013, or they will be chemicals identified as emerging issues.

1.3   Design of Monitoring Programmes in the EU

1.3.1   General
The surface water monitoring network and the respective programmes have been established in all European member states in accordance with the requirements of Article 8 of the Water Framework Directive (WFD). The monitoring network has been designed so as to provide a coherent and comprehensive overview of ecological and chemical status within each river basin. On the basis of the characterisation and impact assessment carried out in accordance with Article 5 and Annex II of the WFD, Member States shall establish for each river basin management plan period three types of monitoring programmes:

- Surveillance monitoring programme,
- Operational monitoring programme,
- And, if necessary, an investigative monitoring programme.

1.3.2   Design of Surveillance and Operational Monitoring Programmes
All available information about chemical pressures and impacts are used for setting up the monitoring strategy. Such information includes substance properties, pressure and impact assessments and additional information on sources, *e.g.* emission data, data on where and for what a substance is used, and existing monitoring data collected in the past.

In many cases, a stepwise screening approach is used to identify non-problem areas, problem areas, major sources, *etc.* This approach starts, for instance, with providing an overview of expected hot spots and sources to gain a first impression of the scale of the problem. Thereafter, a more focused monitoring can be performed, directed to relevant problem areas and sites. For many substances screening of the levels in water as well as in biota with limited mobility and in sediment will be the best way to get the optimum information within a given amount of resources. When problem areas are identified, analysis of a limited number of water samples can be performed.

The monitoring programmes need to take account of variability in time and space (including depth) within a water body. Sufficient samples should be taken and analysed to adequately characterise such variability and to generate meaningful results with proper confidence.

The use of numerical models with a sufficient level of confidence and precision for designing the monitoring programmes can also be helpful. The documentation of progressive reduction in concentrations of priority substances and other pollutants, and the principle of no deterioration, are key elements of WFD and require appropriate trend monitoring. The member states considered this when designing their monitoring programmes. Data obtained in surveillance and operational monitoring are often used for this purpose.

Considering these general aspects, the Irish Environmental Protection Agency (EPA) amongst all other Environment agencies in Europe has set up a surveillance monitoring programme in 2007. The main results of the three years' programme are described in section 3 of this chapter.

1.3.3 Sampling Strategy

Important principles of sampling strategy have been described in relevant harmonised guidance documents. Depending on the objective of the monitoring, the physicochemical properties of the substance to be monitored and the properties of the water body under study, water, sediment and/or biota samples have to be taken. The set-up of the monitoring strategy includes decisions on the sampling locations, sampling frequencies and methods. This selection is commonly a compromise between a sufficient coverage of samples in time and space to generate meaningful results with proper confidence and limiting the monitoring costs.

As the establishment of Environmental Quality Standards (EQS) has been limited for the majority of priority substances to water only, the principle matrix for assessing compliance with respect to EQS is whole water, or for metals, the liquid fraction obtained by filtration of the whole water sample. However, in the case of mercury, hexachlorobenzene and hexachlorobutadiene EQS for biota also have been defined, as outlined in section 1.2 of this chapter. Furthermore, Member States may opt to establish and apply EQSs for sediment and/or biota for other substances listed in the proposed Directive. These EQSs shall offer at least the same level of protection as the EQS for water. For other pollutants, the matrix for analysis should be in line with the matrix for which national EQS have been derived.

1.3.4 Sampling of Water and Suspended Particulate Matter (SPM)

According to the WFD, the chemical status of a water body is generally assessed from analyses of water samples for substances with stated chemical water quality criteria. However, supporting parameters for the assessments of the ecological and chemical status may have to be analysed in water or other matrices.

The type of water sample to be taken at each site is part of the strategy for the monitoring programme. For most water bodies, spot samples are likely to be appropriate. In specific situations, where pollutant concentrations are heavily influenced by flow conditions and temporal variation and if pollution load assessments are to be performed, other more representative types of samples may be beneficial. Flow-proportional or time-proportional samples may be better in such cases. In stratified water bodies, such as lakes, some estuaries and coastal areas, waters samples may be taken in different depths to give a better representation of the water column compared to a single sampling depth. For example, multiparameter probes [*e.g.* Conductivity-Temperature-Depth Probes (CTD probes)] can be employed to detect stratifications.

In general, reliable data on emission sources reduce monitoring costs, because they give a good basis for choosing proper sampling locations, and for optimising the number of sampling sites and the appropriate sampling frequencies. Whole water data may be generated by analysis of the whole water sample or by separate determinations on liquid and SPM fractions. If it can be justified, for example by considerations of expected phase-partitioning of the contaminant, it may be argued that there is not a need to analyse a particular fraction. If a sampling strategy is selected involving only liquid or SPM fractions, then the Member States shall justify the choice with measurements, calculations, *etc.* However, demonstrating compliance with EQS in water may be problematic in some cases.

1.4   Frequency of the Monitoring

The monitoring frequencies, given in WFD, Annex V 1.3.4, of once-a-month for priority substances or once-per-three-months for other pollutants, will result in a certain confidence and precision. More frequent sampling may be necessary, *e.g.* to detect long-term changes, to estimate pollution load and to achieve acceptable levels of confidence and precision in assessing the status of water bodies. In general, it is advisable to take samples in equidistant time intervals over a year, *e.g.* every four weeks, resulting in 13 samples to compensate for missing data due to unusual weather conditions (drought, floods, *etc.*) or laboratory problems.

In the case of pesticides and other seasonally variable substances, which show peak concentrations within short time periods, enhanced sampling frequency compared to that specified in the WFD may be necessary in these periods. For example, the best sampling time for detecting concentration peaks of pesticides due to inappropriate application is after heavy rainfall within or just after the application period. Moreover, failure to comply with good agricultural practice, *e.g.* inappropriate cleaning of equipment during or at the end of the season before winter, can also cause pesticide peak concentrations. Other reasons for enhanced sampling frequency include seasonal pressure from tourism or seasonal industrial activities, which are common practice, for example, in pesticide production, *etc.* The results of those measurements should be compared with the MAC-EQS.

For the calculation of the annual average concentrations, results have to be weighted according to the associated time interval (time-weighted average). For example, 12 equidistant values per year with two additional values in November could be accounted for, with reduced weights for the three November values. In other words, the three November values would be averaged and a "November mean" used in the calculation of the annual average value. Using this approach, any individual values should still trigger an immediate investigation if high levels are detected.

Collecting composite samples (from 24 h up to one week) might be another option to detect peak concentrations of seasonally variable compounds.

Reduced monitoring frequencies and, under certain circumstances, even no monitoring may be justified when monitoring reveals/has revealed that concentrations of substances are far below the EQS, declining or stable, and there is no obvious risk of increase.

The monitoring frequencies quoted in the Directive may not be practical for transitional and coastal waters, Nordic lakes, which can be iced for several months, or for Mediterranean rivers which may contain no water for several months each year.

2   Analytical Challenges of the WFD Monitoring

2.1   Analytical Methods for the Determination of Priority Substances in Water

Article 8 (Paragraph 3) of the WFD requires that technical specifications and standardised methods for analysis and monitoring of water status shall be established in accordance with the procedure laid down in Article 21. Moreover, Annex V.1.3.6 states that the standards for monitoring of quality elements for physicochemical parameters shall be any relevant CEN/ISO standards or such other national or international standards, which will ensure the provision of data of an equivalent scientific quality and comparability.

The strengths of such methods are:

1. They have been developed in a formalised process involving recognised experts from numerous countries;
2. They are well-established, widely accepted and easily available;
3. They mostly have been subjected to collaborative trials to demonstrate their interlaboratory comparability and applicability;
4. There is little effort necessary to implement them as only verification and no full validation is required; and
5. It is possible to refer to them in legislation.

On the other hand, standard methods also feature certain drawbacks, which should be considered when selecting a method for a particular purpose. These include:

1. Standardisation takes quite a long time and hence such methods may not always represent the current state of the art;
2. They usually represent a compromise in performance that is tailored to a number of different users' goals and operational needs and have not been developed specifically for WFD monitoring;
3. Normally they offer little flexibility to the user to choose from different options depending on the nature of samples and so on, *e.g.* what equipment and experience are on hand; and
4. They do not take account of recent technical developments.

For this reason, the majority of Member States expressed concern that requiring the use of specific standardised methods, except for operationally defined parameters, in an EU legal act would not cope with the monitoring requirements set out in the WFD. A different and performance-based approach was therefore suggested. Anyway, the use of standard methods in routine monitoring is widespread across Europe, although the procedures have often been slightly modified and adjusted to the specific measurement requirements, to the nature of the sample, or according to the available equipment and the existing know-how. Standards are often used as reference methods or as vantage point for further method development.

In summary, methods applied for WFD monitoring should clearly be described, properly validated and give laboratories the flexibility to select from several options when possible and meaningful. Irrespective of what method is applied in chemical monitoring, certain minimum performance criteria have to be met. In other words, any fully validated method meeting those criteria may be used. The agreement of Member States as regards the necessary minimum performance criteria of the analytical methods for the WFD monitoring programmes were finally established in the Directive 2009/90/EC, which lays down technical specifications for chemical analysis and monitoring of water status (so called "QA/QC directive").[7]

2.2   The EU QA/QC Directive 2009/90/EC

The Directive 2009/90/EC[7] lays down important technical specifications for chemical analysis and monitoring of the water status. Among other specifications, requirements for the validation of methods, including minimum performance criteria, are fixed. This means that all methods of analysis applied by Member States for the purposes of chemical monitoring programmes of water status have to meet certain minimum performance criteria, including:

1. Rules on the uncertainty of measurements, and on the
2. Limit of quantification of the methods.

To ensure the comparability of chemical monitoring results, the limit of quantification should be determined in accordance with a commonly agreed definition. Where there are no methods which comply with the fixed minimum performance criteria, monitoring should be based on best available techniques not entailing excessive costs.

Member States shall ensure that the minimum performance criteria for all methods of analysis applied are:

1. Based on an uncertainty of measurement of 50% or below (extension factor, $k = 2$) estimated at the level of relevant environmental quality standards; and
2. A limit of quantification equal or below a value of 30% of the relevant environmental quality standards.

2.3 Priority Substances Difficult to Analyse

2.3.1 Organochlorine Pesticides

In the daughter directives of the WFD, EQSs were set for a number of pesticides, including alachlor, atrazine, simazine, diuron, isoproturon, chlorfenvinphos, chlorpyrifos, endosulfan, trifluralin, hexachlorocyclohexanes (HCH), DDT, aldrin, dieldrin, endrin and isodrin.

The most widely used methods for the analysis of pesticides and their metabolites in environmental samples are based on GC-MS and tandem triple-quadrupole LC-MS^2. GC-MS is usually operated in the selected ion monitoring (SIM) and LC-MS^2 in the multiple reaction monitoring (MRM) mode by detecting specific MS/MS mass transitions of the molecules, resulting in increased selectivity. Isotope dilution, based on the addition of labelled internal standards prior to sample extraction, has proven to be a powerful quantitative analytical technique.

Regarding the extraction of water samples, the most commonly used methods are solid-phase extraction (SPE) and liquid–liquid extraction (LLE). Solid-phase micro-extraction (SPME) is less frequently applied. SPE methods are rapid, efficient (good recoveries and low detection limits), use less solvent than LLE and consequently reduce laboratory expenses and waste. In addition, SPE methods can be automated by using laboratory robotic systems that do all or part of the sample preparation steps. SPE also can be combined on-line with LC-MS^2 detection. In addition, highly sophisticated techniques such as triple-quadrupole tandem GC-MS/MS and LC-Q-TOF-MS methods have been reported for pesticide measurements in water.

The more polar triazines (atrazine, simazine) and phenylurea herbicides (diuron, isoproturon), and the chloroacetanilide alachlor, can be analysed by LC-MS, and the other WFD relevant pesticides chlorfenvinphos, chlorpyrifos, endosulfan, trifluralin, hexachlorocyclohexane (HCH), DDT, aldrin, dieldrin, endrin and isodrin by GC-MS only. The method detection limits (MDLs) for the multi-compound pesticide methods based on SPE or LLE followed by GC-MS or LC-MS^2 are usually in the range of 1–15 ng l^{-1} and thus suffice to meet the requirements of the WFD in most cases.

However, AA-EQSs in particular those for other surface waters as defined in Sec. 1.2 for endosulfan, hexachlorocylohexane and the sum of cyclodiene pesticides, aldrin, dieldrin, endrin and isodrin, are significantly lower than the typical limit of quantification (LOQ) of routinely applied analytical methods. For example, endosulfan has a relatively low sensitivity in GC-MS analysis, with method detection limits of about 10 ng l^{-1}. Lower LOQ for chlorinated pesticides might be achieved by extracting large sample volumes (>1 litre), applying large-volume injection or using GC-ENCI-MS, respectively, which offers higher sensitivity compared to electron ionisation.

There is a need to validate those methods by proper in-house validation studies and to organise inter-laboratory comparison on the determination of WFD-relevant pesticides in natural water samples. SPE-LC-MS^2 appears to be the method of choice to control the quality standards for individual pesticides and the sum of all individual pesticides, including their relevant metabolites, degradation and reaction products of 0.1 and 0.5 µg l^{-1}, respectively, set out in the Directive on the protection of groundwater against pollution and deterioration. Pesticides without an aromatic structure, like glyphosate (currently the most widely used herbicide) and its metabolites AMPA and glufosinate, can only be analysed by GC-MS or LC-MS methods after derivatisation.

2.3.2   Polycyclic Aromatic Hydrocarbons (PAHs)

The Decision 2455/2001/EC2, establishing a list of priority substances for European water policy, includes eight PAHs of which six (anthracene, benzo[*a*]pyrene, benzo[*b*]fluoranthene, benzo[*g,h,i*]perylene, benzo[*k*]fluoranthene and indeno[*1,2,3-c,d*]pyrene) are classified as priority hazardous substances, and naphthalene as well as fluoranthene as priority substances, respectively.[2] International standardised methods for analysing PAHs in surface waters are available, including ISO 17993 based on high-performance liquid chromatography (HPLC) with fluorescence detection after liquid–liquid extraction and the new draft standard method (ISO/DIS 28540) based on gas chromatography-mass spectrometry (GC-MS) after either liquid–liquid or liquid–solid disk extraction. Moreover, EPA Method 525, employing C 18 cartridges or C 18 disks for PAH extraction, is widely used as well. While various methods for PAH analysis in water exist, the following analytical challenges for compliance monitoring under the WFD remain:

1. Requirements on method sensitivity derived from current EQSs for the sums of benzo[*b*]fluoranthene and benzo[*k*]fluoranthene (0.03 µg l^{-1}), as well as for benzo[*g,h,i*]perylene and indeno[*1,2,3-c,d*]pyrene) (0.002 µg l^{-1}), respectively, are lower than the LOQs of the two cited ISO methods (0.01 and 0.005 µg l^{-1}) and, hence, difficult to meet.
2. EQSs refer to whole water concentrations, *i.e.* the sum of dissolved and particle-bound PAH concentrations is to be reported, irrespective of whether derived from whole water analysis or from separate analyses of the liquid and particulate fractions.

An overview of the existing approaches shows that, in particular, methodologies based on simultaneous extraction and analysis of dissolved and particulate-bound PAHs, as proposed in the draft ISO standard method ISO/DIS 28540, would have the potential to be applied for cost-effective routine WFD monitoring if method sensitivity could be improved towards required levels. Furthermore, extraction efficiencies for PAHs from SPM, as well as the robustness of the proposed techniques, need to be elaborated under routine conditions, including samples containing up to 500 mg l^{-1} of SPM.

2.3.3   Tributyltin Compounds

The AA-EQS for tributyltin is significantly lower than the typical limits of quantitation (LOQ) for routinely applied analytical methods. The AA-EQS of

$0.2\,\mathrm{ng}\,\mathrm{l}^{-1}$ suggests a required LOQ of $0.06\,\mathrm{ng}\,\mathrm{l}^{-1}$, whereas current commonly used analytical methodologies have typical LOQs in the $1–10\,\mathrm{ng}\,\mathrm{l}^{-1}$ range. Therefore, there is a need for further research to either improve the sensitivity of existing techniques, develop new practical methodologies or, perhaps, to utilise a compliance regime based on the analysis of a suitable biological receptor and take advantage of the associated biomagnification. There is a wide variety of different analytical techniques that have been employed for the measurement of organotin species in environmental compartments. The diversity of different techniques that have been employed over recent years for the analysis of tributyltin probably reflects the recognition of it being a significant pollutant. Analytical methodologies for tributyltin are well represented in the literature and have been reviewed.[8,9]

While there are many potential techniques for the determination of tributyltin in waters, the majority of routine methods employed for environmental monitoring usually require the derivatisation of tributyltin to a more volatile alkyltributyltin or hydride, prior to measurement by gas chromatography (GC) with either mass spectrometry (MS), flame photometric detector (FPD), or occasionally atomic emission detector (AED). The butyltin hydrides tend to be unstable when stored for any length of time, which can, for example, limit the ability to re-analyse extracts. For this reason, alkylation tends to be the preferred derivatisation process, as the alkylated derivatives are relatively stable. However, alkylation is not the only option for derivatising organotins; forming 4-fluorophenyl derivatives has been reported to offer cleaner chromatograms[10] and, therefore, the opportunity for improved LOQ.

Using modern sensitive GC-MS or GC-FPD instruments, together with optimisation of the sample inlet conditions using, for example, large-volume injection, can reduce the limit of detection (LOD) for tributyltin down to $0.5\,\mathrm{ng}\,\mathrm{l}^{-1}$ or less. However, the LOD is often not limited by sensitivity alone, as interfering peaks in chromatograms or positive and variable blanks can often raise the true LOD.

Isotope dilution techniques offer excellent accuracy and precision compared to most other measurement methods. Once viewed perhaps as an exotic technique best suited to research laboratories only, the commercial availability of tin isotope-enriched butyltins and its increased use as a technique for organometal analysis have brought this technique potentially within the reach of routine laboratories. Isotope dilution has been undertaken using a chromatographic delivery system (GC or HPLC) linked to an inductively coupled plasma mass spectrometer (ICP-MS). However, ICP-MS instruments are relatively expensive and this could limit the routine use of this technique. Alternatively, isotope dilution analysis has been undertaken by GC-MS. While this technique would probably be suitable for routine use, the literature again acknowledges the need for especially clean sample preparation conditions in order to achieve the best method performance.

In summary, while analytical methods exist for tributyltin that could potentially achieve the required LOQ (at least in seawater), there is a need for further research to produce a method that can accommodate the variety of waters to be monitored for the WFD and that can also be used in routine laboratory conditions.

2.3.4   Pentabromodiphenylether (PBDE)

The proposed AA-EQSs for the sum of BDE28, BDE47, BDE99, BDE100, BDE153 and BDE154 in inland and other surface waters (transitional, coastal and territorial waters) are 0.5 and $0.2 \, \text{ng} \, l^{-1}$, respectively. The LOQ of analytical methods used for compliance checking should be below 30% of these values. If a 1 litre water sample is extracted, as it is common practice in many monitoring laboratories, the required low LOQ can, presumably, not be achieved. Furthermore, it has to be considered that, depending on the content and the characteristics of suspended particulate matter (SPM) present in the sample, the dissolved PBDE levels may vary significantly. Hence, both fractions (particle-bound and dissolved) have to be taken into account. Therefore, there is a need for further research to validate extraction procedures with regard to their applicability to whole water samples and to develop new practical enrichment techniques to improve sensitivity.

Currently, no European and international standard methods for the analysis of PBDEs in water exist, although such analyses are performed in research and contract laboratories. There is an ISO standard method for the analysis of PBDEs in sediment and sewage sludge using GC-MS in the electron ionisation or ECNI mode (ISO 22032). Specifications and experimental conditions given in this ISO method might also be useful for measurements of PBDEs in water.

2.3.5   Short-Chain Chlorinated Paraffins (SCCPs)

An accurate chemical analysis of SCCPs in environmental samples is difficult to achieve due to the highly complex nature of commercial formulations, the impact of numerous physical, chemical and biological processes after use, and the lack of certified reference standards. SCCPs are very complex mixtures containing many congener groups chlorinated to various degrees and at different positions on the carbon backbone (about 7800 isomers with a carbon skeleton between C10 and C13). Comprehensive overviews of the analytical methodologies applied to the analysis of chlorinated paraffins have been provided in recently published reviews.[11-14]

In brief, there is a variety of approaches to analyse SCCPs in environmental samples, based on capillary gas chromatography (GC) in combination with various detectors:

1. GC-electron capture detection,
2. GC-electron capture negative ionisation-mass spectrometry (ECNI-MS),
3. GC-dichloromethane enhanced negative ion chemical ionisation-mass spectrometry,
4. GC-electron ionisation mass spectrometry,

5. GC-metastable atom bombardment ionisation (MAB) high-resolution mass spectrometry, and
6. Carbon skeleton analysis, after simultaneous catalytic dechlorination and hydrogenation, by gas chromatography with flame ionisation or mass spectrometric detection.

Furthermore, considerable improvement in the separation of SCCPs was recently achieved by using comprehensive two-dimensional gas chromatography (GC×GC) coupled to a rapid-scanning quadrupole mass spectrometer (qMS) or time-of-flight mass spectrometer (TOF-MS), both operated in the electron-capture negative-ion mode. Currently, GC-ECNI-MS is the most widely used technique to analyse SCCPs in environmental samples. Drawbacks are the discrimination of isomers with 3–5 chlorine atoms and the large differences in response factors, depending on the number of chlorine atoms in the molecules. The latter has some implication as regards the selection of an appropriate calibration standard and quantification. Until recently, technical or synthetic mixtures with known chlorine-content have been used for calibration purposes. An international inter-laboratory study indicated that some of the observed variability in the analytical results may be introduced when different commercial formulations are used as external standards. These findings emphasise the importance of the choice of suitable standards for quantitative analysis, which should match the SCCP pattern in the sample as far as possible.

Pellizzato *et al.*[15] proposed the definition of an operationally defined parameter as surrogate for SCCPs on the basis of carbon skeleton analysis after simultaneous catalytic dechlorination and hydrogenation by gas chromatography with flame ionisation or mass spectrometric detection. That means that the determinand is defined *via* the application of a precisely described analytical procedure, which provides also the reference for the metrological traceability of the measurement results. Such a procedure would simplify analysis of SCCPs considerably; however, information on the degree of chlorination gets lost. This information, although important from a toxicological point of view, is not needed to meet the monitoring requirements of the WFD.

While some work has been conducted on development of sensitive methods for SCCP analysis in recent years, currently no fully validated procedure is available that could be recommended for routine monitoring of SCCPs in water for the purpose of compliance checking. Given that Member States are obliged to report SCCP results, it can be assumed that unacceptable uncertainty will be associated with these results and their reliability and comparability must be questioned.

For this reason, there is an urgent need to develop and validate methods to enable Member States to provide reliable data on total-SCCP content in European waters and thus to meet the legal requirements set out in the WFD. Method development should rather be directed towards a simple total-SCCP approach than a formula group approach. Furthermore, to comply with European legislation proper quality assurance tools have to be provided. In this respect, the production of standard reference materials for calibration purposes, isotopically labelled reference standards and matrix reference materials certified for their SCCP content is required, and inter-laboratory exercises need to be conducted.

3   Case Study: Surface Water Monitoring in Ireland

3.1   Introduction

Between July 2007 and December 2009, the chemical monitoring programme for surface waters in the Republic of Ireland was put into place. A total number of about 70 lake sites and 180 river sites were selected for chemical monitoring according to the requirements of the WFD. Sampling was carried out on a monthly basis and a number of physicochemical parameters were analysed in each sample. In each year, a sub-set of about a third of the sampling sites was selected for monthly analysis of "priority action substances" (PAS).

This parameter suite comprised all substances listed in Annex 1 of the WFD daughter directive on Environmental Quality Standards.[4] This annex includes all 33 priority and priority hazardous substances of the WFD and certain other pollutants for which environmental quality standards (EQS) have been defined at the European level. Furthermore, the Republic of Ireland decided to include several other pollutants of potential local relevance in this monitoring programme and defined provisional national EQS values for some of these additional parameters.[16]

In the following, the term "priority action substances" (PAS) is used for this suite of parameters derived from the WFD, its daughter directives and national Irish regulations.

In total, 8216 samples were taken and 2502 samples were analysed for the complete PAS suite during the 30-month monitoring period. A list of the parameters analysed is given in Table 1.

3.2   Overview of Results of the Chemical Monitoring of Priority Substances

3.2.1   Substances with Concentrations below LOQ

For 24 of the parameters (and parameter groups) that were analysed, results were always below the limit of quantification (LOQ). These parameters are marked by putting the number of the substance in parentheses (first column in Table 1). However, it should be considered that for 8 of these 24 parameters, the analytical methods used were not sensitive enough to check compliance with the European or national Environmental Quality Standards (EQS). In particular, analytical methods for the following parameters that were never detected at quantifiable levels were not sufficiently sensitive to quantify concentrations at or below the EQS (*cf*. sections 2.1 and 2.3).

- Pentabromodiphenylethers
- C10–C13 chloroalkanes
- Endosulfan

Table 1 Priority action substances (PAS) monitored in surface waters of Ireland 2007–2009. Numbering in first column refers to EQS directive,[2] substances with prefix **IRL** in numbering were added at the national level.

Number[a]	Substance name	CAS Number	EQS[b] in µg/l	Analysis technique	LOQ[c] in µg/l
(1)	Alachlor	159772-60-8	0.3	EN ISO 11369	0.01
2	Anthracene	120-12-7	0.1	EN ISO 17993	0.005
3	Atrazine	1912-24-9	0.6	EN ISO 10695	0.01
4	Benzene	71-43-2	10	GC-MS	
(5)	Pentabromodiphenylether (PBDE)[d]	32534-81-9	0.0005	ISO/DIS 22032	0.2
6	Cadmium and its compounds	7440-43-9	0.08–0.25[e]	ISO 17294-2	
6a	Carbon tetrachloride	56-23-5	12		
(7)	C10-13-chloroalkanes	85535-84-8	0.4	SPE-GC-NCI-MS	5
(8)	Chlorfenvinphos	470-90-6	0.1	EN 12918	0.02
9	Chlorpyrifos-ethyl	2921-88-2	0.03	EN ISO 10695	0.02
(9a)	Cyclodiene pesticides:		$\sum = 0.010$		
	Aldrin	309-00-2		EN ISO 6468	0.01
	Endrin	60-57-1		EN ISO 6468	0.01
	Dieldrin	72-20-8		EN ISO 6468	0.01
	Isodrin	465-73-6		EN ISO 6468	0.01
(9b)	DDT total[f]	n/a	0.025	EN ISO 6468	0.01
	p,p'-DDT	50-29-3	0.01	EN ISO 6468	0.01
(10)	1,2-Dichloroethane	107-06-2	10	GC-MS	
11	Dichloromethane	75-09-2	20	GC-MS	
12	Di(2-ethylhexyl)phthalate (DEHP)	117-81-7	1.3	EN ISO 18856	0.05
13	Diuron	330-54-1	0.2	EN ISO 11369	0.03
(14)	Endosulfan	115-29-7	0.005	EN ISO 6468	0.01
15	Fluoranthene	206-44-0	0.1	EN ISO 17993	0.005
(16)	Hexachlorobenzene	118-74-1	0.01	EN ISO 6468	0.01
(17)	Hexachlorobutadiene	87-68-3	0.1	EN ISO 6468	0.01
(18)	Hexachlorocyclohexane (HCH)[g]	608-73-1	0.02	EN ISO 6468	0.01
19	Isoproturon	34123-59-6	0.3	EN ISO 11369	0.03
20	Lead and its compounds	7439-92-1	7.2	ISO 17294-2	1

Table 1 Continued.

Number[a]	Substance name	CAS Number	EQS[b] in μg/l	Analysis technique	LOQ[c] in μg/l
(21)	Mercury and its compounds	7439-97-6	0.05	ISO 17294	0.01
22	Naphthalene	91-20-3	2.4	GC-MS	0.1
23	Nickel and its compounds	7440-02-0	20	ISO 17294	1
(24)	4-Nonylphenol	25154-52-3	0.3	ISO 18857-2	0.02
(25)	Octylphenol (para-tert-Octylphenol)	140-66-9	0.1	ISO 18857-2	0.02
(26)	Pentachlorobenzene	608-93-5	0.007	EN ISO 6468	0.01
27	Pentachlorophenol	87-86-5	0.4	EN ISO 15913	0.01
28	Polyaromatic Hydrocarbons (PAH)	n/a	n/a	EN ISO 17993	n/a
	Benzo(a)pyrene	50-32-8	0.05	EN ISO 17993	0.002
	Benzo(b)fluoroanthene	205-99-2	$\sum = 0.03$	EN ISO 17993	0.005
	Benzo(k)fluoroanthene	207-08-9		EN ISO 17993	0.005
	Benzo(g,h,i)perylene	191-24-2	$\sum = 0.002$	EN ISO 17993	0.005
	Indeno(1,2,3-cd)-pyrene	191-39-5		EN ISO 17993	0.005
29	Simazine	122-34-9	1	EN ISO 10695	0.01
29a	Tetrachoro-ethylene	127-18-4	10	GC-MS	
29b	Trichloro-ethylene	79-01-6	10	GC-MS	
(30)	Tributyltin compounds (TBT cation)	688-73-3	0.0002	GC-AED	0.001
31	Trichloro-benzenes	12002-48-1	0.4	GC-MS	
32	Trichloro-methane	67-66-3	2.5	GC-MS	
(33)	Trifluralin	1582-09-8	0.03	EN ISO 10695	0.01
(IRL01)	Epichlorohydrin	106-89-8	0.1	EN 14207	0.05
IRL02	Mecoprop	96-65-2	0.02	EN ISO 15913	0.02

Code	Substance	CAS	Value	Method	Value
(IRL03)	Pirimiphos-methyl	29232-93-7	0.05	EN ISO 10695	0.02
(IRL04)	Fenitrothion	122-14-5	0.01	EN ISO 10695	0.02
(IRL05)	Malathion	121-75-5	0.01	EN 12918	0.08
IRL06	Epoxiconazole	135319-73-2	0.1	EN ISO 10695	0.01
IRL07	Glyphosate	1071-83-6	0.1	HPLG'	0.08
(IRL08)	Nonylphenol ethoxylates	37340-60-6	0.1	ISO 18857-2	0.01
(IRL09)	Dithiocarbamate pesticides:				
	Maneb	124727-38-2	0.1	Sum parameter: detection of degradation product CS_2	$\Sigma = 10$
	Thiram	137-26-8	0.1		
	Mancozeb	8018-01-7	0.1		
	Zineb	12122-67-7	0.1		
IRL10	Cyanide		10	EN ISO 14403	2
IRL11	Dichlobenil	1194-65-6	0.1	EN ISO 6468	0.04
IRL12	2,6-dichlorobenzamide	2008-58-4	0.1	EN ISO 10695	0.06
IRL13	Linuron	330-55-2	0.1	EN ISO 10695	0.04
IRL14	2,4-D	94-75-7	0.1	EN ISO 15913	0.01

[a] According to the EQS directive.[2] If the number is given in brackets, this substance has not been detected above the LOQ in any sample during the whole monitoring campaign.
[b] Environmental Quality Standard expressed as an annual average for inland surface waters.
[c] LOQ: Limit of Quantification
[d] Comprises the sum of the congeners No 28, 47, 99, 100, 153 and 154.
[e] Depending on water hardness (5 classes).
[f] DDT total comprises 4 isomers: op'-DDT, pp'-DDT, pp'-DDD, pp'-DDE.
[g] HCH includes the isomers alpha-HCH, beta-HCH, gamma-HCH (Lindane), delta-HCH.

- Pentachlorobenzene
- Tributyltin compounds
- Fenitrothion
- Malathion
- Dithiocarbamate pesticides.

3.2.2 *Substances with Concentrations above LOQ (Positive Results)*

Positive results (*i.e.* concentrations equal to or above the LOQ) were obtained for 31 of the monitored parameters. A ranking of these parameters according to their occurrence is given in Table 2. Only six of these 31 parameters were detected in more than 10% of the samples:

- Nickel
- Di(2-ethylhexyl)phthalate (DEHP)
- Benzo[*a*]pyrene
- Fluoranthene
- Trichloromethane
- Cyanide.

These six substances are widespread over a wide range of water bodies and were detected at about 50% of the sampling sites (at least once during the monitoring campaign). The highest occurrence in terms of geographical spread has been observed for DEHP, which has been detected at 77% of the sampling sites in at least one sample.

However, before drawing conclusions about the contamination level of water bodies by these substances, it should be considered that, for most of these parameters, very sensitive methods were used with LOQs usually in the range of 5% of the EQS (except for cyanide where the LOQ is 20% of the national EQS). In spite of the high occurrence, concentration levels of these substances were usually far below the respective EQS (see Table 1).

3.2.3 *Substances with Concentrations above the EQS*

Only a subgroup of 17 of the 31 detectable parameters was ever measured at concentration levels equal to or above the respective national or European EQS. A ranking of all quantifiable parameters according to the frequency of EQS exceedances is given in Table 3. The highest ranking parameters according to EQS exceedances are not identical with the parameters with the highest occurrence as given in Table 2.

There are only six parameters that are present in concentrations above the EQS in ten or more samples:

- Mecoprop
- Glyphosate
- Sum of Benzo[*ghi*]perylene and Indeno[*1.2.3-cd*]pyrene
- Sum of Benzo[*b*]fluoranthene and Benzo[*k*]fluoranthene

- Benzo[*a*]pyrene
- Fluoranthene.

For most substances or substance classes, samples with concentrations ≥ EQS were usually singular events and did not indicate a constant contamination of the investigated water bodies. However, in case of mecoprop, glyphosate, and PAH parameters such as "sum of benzo[*g,h,i*]perylene and indeno[*1,2,3-c,d*]pyrene" and "sum of benzo[*b*]fluoranthene and benzo[*k*]-fluoranthene", several sites showed repeatedly concentrations ≥ EQS, indicating that an input of these substances is a more constant pressure on surface waters at some locations. This can be seen from the differences between the

Table 2 Ranking of detected PAS according to occurrence – number and percentage of samples and sites where substance where detected at concentration levels above LOQ.

Parameter name	No. of samples	No. of sites	Samples in %	Sites in %
Nickel	761	127	30	50
Di(2-ethylhexyl)phthalate	463	195	19	77
Benzo[a]pyrene	388	137	16	54
Fluoranthene	326	126	13	50
Trichloromethane	307	121	12	48
Cyanide	258	138	10	54
Simazin	232	93	9.3	37
Mecoprop	195	72	7.8	28
Benzo[b]fluoranthene & Benzo[k]fluoranthene	159	77	6.4	30
Glyphosate	144	81	5.8	32
Atrazine	131	68	5.2	27
Benzo[ghi]perylene & Indeno[1.2.3-cd]pyrene	120	70	4.8	28
Benzene	102	66	4.1	26
Naphthalene	93	80	3.7	31
Lead	71	11	2.8	4
2.4-Dichlorophenoxyacetic acid	66	48	2.6	19
Isoproturon	56	28	2.2	11
Diuron	42	29	1.7	11
Dichloromethane	39	34	1.6	13
Trichlorobenzenes	25	22	1.0	9
Tetrachloroethylene	19	14	0.8	6
Trichloroethylene	17	6	0.7	2
Cadmium	15	4	0.6	2
Dichlobenil	11	7	0.4	3
Anthracene	2	1	<0.1	0.4
Epoxiconazole	2	2	<0.1	0.8
2.6-Dichlorobenzamide	2	2	<0.1	0.8
Carbon tetrachloride	2	2	<0.1	0.8
Chlorpyrifos-ethyl	1	1	<0.1	0.4
Linuron	1	1	<0.1	0.4
Pentachlorophenol	1	1	<0.1	0.4

Table 3   Ranking of detected PAS according to exceedance of European or national or provisional EQS values. Number and percentage of sites and samples where concentration of PAS was equal to or above the EQS. PAS that were never detected in any sample are not listed.

Parameter name	No. of samples	No. of sites	Samples in %	Sites in %
Mecoprop	195	72	7.8	28
Glyphosate	131	78	5.2	31
Benzo[ghi]perylene & Indeno[1.2.3-cd]pyrene	120	70	4.8	28
Benzo[b]fluoranthene & Benzo[k]fluoranthene	35	25	1.4	10
Benzo[a]pyrene	11	9	0.4	3.5
Fluoranthene	10	8	0.4	3.2
2.4-Dichlorophenoxyacetic acid	10	10	0.4	3.9
Isoproturon	8	5	0.3	2.0
Dichlobenil	6	6	0.2	2.4
Trichloromethane	4	4	0.2	1.6
Trichlorobenzenes	4	4	0.2	1.6
Di(2-ethylhexyl)phthalate	3	3	0.1	1.2
Diuron	3	3	0.1	1.2
Naphthalene	1	1	< 0.1	0.4
Lead	1	1	< 0.1	0.4
2.6-Dichlorobenzamide	1	1	< 0.1	0.4
Chlorpyrifos-ethyl	1	1	< 0.1	0.4
Nickel	0	0	0.0	0.0
Cyanide	0	0	0.0	0.0
Simazin	0	0	0.0	0.0
Atrazine	0	0	0.0	0.0
Benzene	0	0	0.0	0.0
Dichloromethane	0	0	0.0	0.0
Tetrachloroethylene	0	0	0.0	0.0
Trichloroethylene	0	0	0.0	0.0
Cadmium	0	0	0.0	0.0
Anthracene	0	0	0.0	0.0
Epoxiconazole	0	0	0.0	0.0
Carbon tetrachloride	0	0	0.0	0.0
Linuron	0	0	0.0	0.0
Pentachlorophenol	0	0	0.0	0.0

number of samples and the number of sites with EQS exceedances as given in Table 3 for each parameter.

Mecoprop and glyphosate are not regulated by the WFD or its daughter regulations, but had been included in the monitoring programme by the Environmental Protection Agency of Ireland. The EQSs are, therefore, not formal European or National EQSs but provisional target levels for analyses that are probably subject to revision as national EQSs.

There are distinct differences in the seasonal patterns of the pesticides mecoprop and glyphosate, as compared to the PAH. In Figure 1, the number of samples with concentrations of mecoprop and glyphosate is added up per month

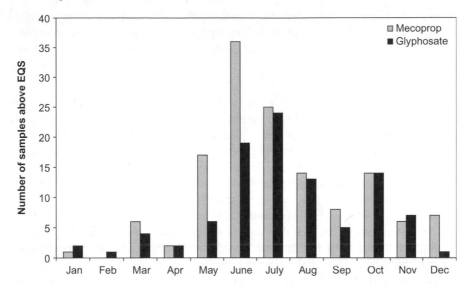

Figure 1   Seasonality of Mecoprop and Glyphosate concentrations ≥ EQS in 2008
and 2009 (cumulative).

(for 2008 and 2009, the two years with a complete annual monitoring cycle),
whereas the monthly aggregation for the two highest ranking PAH parameters
is given in Figure 2. Glyphosate and mecoprop follow a clear seasonal pattern
linked to the growth period (highest occurrence during the summer months),

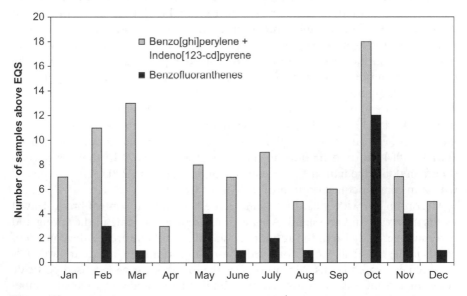

Figure 2   Seasonality of PAH results ≥ EQS in 2008 and 2009 (cumulative).

whereas incidents with elevated PAH concentrations occur more evenly spread throughout the year, with even a slightly higher number in the winter months.

3.3   Discussion

In general, the concentration of most priority substances of the WFD is considerably below the EQS values in the investigated water bodies in Ireland and nearly half of the priority substances were even not detectable at all in any of the samples. There is only a limited number of parameters of the EQS directive that were detected in a small number of samples at concentrations above the EQS. Usually, these were single incidents which will not lead to an exceedance of the EQS defined as an annual average concentration.

However, there are a few parameters that were detected in a higher number of samples; in particular, two parameters which were added to the monitoring programme by the EPA Ireland (mecoprop and glyphosate) with a provisional national EQS.

In addition to these two parameters, most results $\geq$ EQS were observed for parameters belonging to the compound class of polycyclic aromatic hydrocarbons (PAHs). In the following, the results of these parameters will be discussed as regards their relation to seasonality, correlation to application patterns and findings from other countries.

3.3.1   Mecoprop

The maximum observed mecoprop concentration in this study was $3.1\,\mu g\,l^{-1}$, whereas the median of the samples with quantifiable mecoprop concentrations (*i.e.* $\geq 0.02\,\mu g\,l^{-1}$) in the investigated surface waters of Ireland in this study was $0.04\,\mu g\,l^{-1}$. The situation in Ireland (occurrence and EQS exceedance in 7.8% of the sample) seems to be quite comparable to the results reported from Germany: In a recent study, mecoprop was ranked among the 10 most frequently found pesticides in surface waters in Germany, with $0.09\,\mu g\,l^{-1}$ as the median of the reported maximum concentrations from numerous surface waters.[17] For the Rhine river, quantifiable mecoprop concentrations were observed in 24 out of 482 samples (5%) in 2006, with concentrations ranging from 0.01 to a maximum value of $0.14\,\mu g\,l^{-1}$ (ref. 17). In the Ruhr river, quantifiable concentrations of mecoprop were observed in 2–9.6% of investigated annual sample series from 2001 to 2004 (no information on the concentration level). In the Stever river (Germany), the median of the mecoprop concentration in 2005 was $0.03\,\mu g\,l^{-1}$, with a maximum concentration of $0.13\,\mu g\,l^{-1}$ (ref. 17).

Mecoprop is a widely used post-emergence herbicide to control broad-leaved weeds in cereals and grassland. Applications are usually made in the spring and early summer and also in the autumn. It is used in horticulture for the control of weeds under top-fruit crops, for weed control in turf and by amateur gardeners for weed control in lawns. Mecoprop was the fourth most widely used pesticide active ingredient used on arable crops in England and Wales in 1990; an annual application rate in the UK of 4000 tonnes was reported for 1988 (ref. 18).

The widespread occurrence of mecoprop (nearly a third of the sites, *cf.* Table 3) and the distinct seasonal pattern (see Figure 1) observed in this study are therefore in line with the facts on application pattern and rates.

To date, no National Irish EQS has been set for mecoprop but, in order to put the data in context, the United Kingdom Technical Advisory Group for the WFD (UKTAG) have proposed a value of $5.5 \mu g l^{-1}$ for England and Wales. This is above the highest value from this study.

3.3.2 Glyphosate

The wide geographical distribution of glyphosate across a significant number of rivers and lakes, as well as the fading of glyphosate concentrations in late autumn and winter, correlate well with the application patterns of glyphosate. Glyphosate is the active agent in a number of herbicide formulations. It is frequently used in agriculture (*e.g.*, in production of cereals, beet culture, maize production) but also for covered/sealed areas. It is a non-selective, broad-spectrum, post-emergence herbicide that is highly effective against emerged grasses, brush and broad-leaf weeds. Glyphosate is highly soluble in water (11 600 ppm at 25 °C), with an octanol–water coefficient (log$K$ow) of -3.3. It is stable in water at pH 3, 5, 6, and 9 at 35 °C. It is also stable to photodegradation in pH 5, 7 and 9 buffered solutions under natural sunlight. The hydrolysis half-life is >35 days.[19]

The maximum observed glyphosate concentration in this study was $1.8 \mu g l^{-1}$. The median of the quantifiable glyphosate concentrations (*i.e.* $\geq 0.08 \mu g l^{-1}$) in the investigated surface waters of Ireland in this study was $0.17 \mu g l^{-1}$, the arithmetic mean was $0.30 \mu g l^{-1}$. Official data on glyphosate concentrations in surface waters from other European countries are scarce. In a recent study, glyphosate was ranked as the ninth most frequently found pesticide in surface waters in Germany, with $0.10 \mu g l^{-1}$ as the median of the reported maximum concentrations from numerous surface waters.[17] For the Rhine river, quantifiable glyphosate concentrations were observed in 139 of 325 samples (43%) in 2006, with concentrations ranging from 0.01 to a maximum value of $4.83 \mu g l^{-1}$ (ref. 17).

The recent Irish Water Framework Directive Surface Water Regulations have set a National EQS of $65 \mu g l^{-1}$ for glyphosate,[16] which is considerably greater than the highest value from the survey.

3.3.3 Polycyclic Aromatic Hydrocarbons (PAHs)

The other four parameters that were detected at concentrations above the EQS in 10 or more samples (*cf.* section 3.2.3) are priority substances according to the WFD and they all belong to the same compound class of polycyclic aromatic hydrocarbons (PAHs). No apparent geographical or seasonal pattern could be observed. This indicates that events of increased PAH concentrations in the monitored rivers of Ireland do not originate from significant and constant point sources. PAHs are a ubiquitous substance group with a number of diverse sources (natural and anthropogenic). As PAHs have a strong affinity towards the solid phase, their occurrence in whole water samples is often linked to flux conditions that lead to a temporarily increased load of suspended particulate matter (SPM) in the water body. Furthermore, the slightly higher number of samples with PAH concentrations above the EQS during the winter months may be partly due to burning of coal or turf during the heating period and subsequent atmospheric transport and deposition.

3.4   Challenges and Pitfalls

Apart from the logistical and analytical challenges of handling a 30-month monitoring campaign with about 8200 samples, there were two demanding analytical challenges, one related to the analysis of the parameter di(2-ethyl-hexyl)phthalate (DEHP) and another one related to tributyltin. In both cases, there was serious concern that the positive results may, at least partly, be an artefact due to sample contamination in the field or in the laboratory.

3.4.1   Tributyltin

In 2007 and 2008, no samples with quantifiable concentrations of tributyltin (TBT) were found. However, in 2009 TBT was found in 26 samples (lakes only).

As this sudden increase of TBT concentrations was regarded as implausible, a range of additional QA/QC measures and tests were taken in order to check whether these TBT concentrations were actually present in the lake or were an analytical artefact. One of the first measures was the use of field blanks.

TBT-free water was taken to the sampling sites (onto the boat at the lake) and was transferred on-site into sampling bottles at the actual sampling location, simultaneously with the actual sampling of lake water. In all of these samples TBT concentrations were below the detection limit. Therefore, bottles, packaging material, handling of bottles and the laboratory process were ruled out as sources of a TBT contamination.

As a second step, sediment samples were taken from selected lakes where there were positive analytical findings for TBT in the water samples analysed. This was done in order to check whether there were additional indicators for an actual contamination of lakes by TBT. Sediment samples were taken at three different locations of each lake designated to be sampled. In all sediment samples, concentrations of organotin compounds were below detection limit ($<1 \, \mu g \, kg^{-1}$). Therefore, an actual TBT contamination of the lakes was considered unlikely and further effort was taken to identify the source of TBT concentrations in lake samples.

Statistical evaluation and correlation analysis of the positive TBT results indicated that only those lakes were affected where a specific type of inflatable boat was used for sampling. Boats of the same type and from the same manufacturer had been used in the sampling campaigns of 2007 and 2008, without any positive TBT result. In 2009, two new boats of the same type and manufacturer came into use for lake sampling.

Table 4 Organotin compounds in boat painting and water after contact with the boat surface material.

Sample	Monobutyltin (MBT)	Dibutyltin (DBT)	Tributyltin (TBT)	Tetrabutyltin (TTBT)
Boat paint	$3170\,\mu g\,kg^{-1}$	$5200\,\mu g\,kg^{-1}$	$3550\,\mu g\,kg^{-1}$	$173\,\mu g\,kg^{-1}$
Water	$91\,ng\,l^{-1}$	$817\,ng\,l^{-1}$	$1336\,ng\,l^{-1}$	$<1\,ng\,l^{-1}$

In order to clarify whether the positive TBT results might be due to the new sampling boats used in 2009, the boat material was tested. A paint sample was scraped off the transom of the inflatable boat and analysed. In addition, a leaching test of the boat material was carried out. To this purpose, the inside of the inflated boat was rinsed with tap water several times, then filled with tap water to several cm above floor level. The water sample was taken after approximately 10 minutes.

TBT results of these samples indicated that the new boats actually acted as source for the TBT traces in lake samples. As lake samples were usually taken by manually scooping water from around the boat the water close to the boat may have been contaminated by organotin compounds leaching from the coating into the water. Results of organotin analysis of leachate and coating material are presented in Table 4.

Any subsequent sampling of lakes was carried out with different boats (that had been used in 2007 and 2008) and samplers took increased effort in reaching out beyond the immediate surrounding of the boat.

We conclude that positive TBT results in lake samples in 2009 do not indicate an actual contamination of the lakes but were caused by leaching of TBT compounds from the surface coating of inflatable boats used for lake sampling. Therefore, no TBT results were reported for the affected lake samples.

3.4.2 Di(2-ethylhexyl)phthalate (DEHP)

In 463 samples (from 127 sites), concentrations of di(2-ethylhexyl)phthalate (DEHP) were quantifiable. This means that at 50% of the sites there was at least one sampling event with a positive DEHP result. As the analysis of DEHP is known to be very demanding and prone to false negative results due to ubiquitous background contamination,[20,21] additional QA/QC measures were taken in order to ensure that the measured results actually reflected DEHP concentrations in the water bodies and were not analytical artefacts. Similar to the case of TBT, field blanks were taken at those sites where positive DEHP results had occurred repeatedly. Furthermore, all positive DEHP results were checked for whether there was any correlation with sampling equipment, sampling staff, sampling technique, *etc.*

As none of the field blanks gave positive DEHP results and no correlation between positive DEHP results and any sampling factor could be identified, it was concluded that DEHP results reflect the actual concentration level at the monitored sites.

References

1. Directive 2000/60/EC of 23 October 2000 establishing a framework for Community action in the field of water policy, *Off. J. Eur. Comm.*, **L327**, 22.12.2000, 1.
2. Decision 2455/2001/EC of 20 November 2001 establishing a list of priority substances in the field of water policy, *Off. J. Eur. Comm.*, **L331**, 15.12.2001, 1.
3. *Guidance on Surface Water Chemical Monitoring under the Water Framework Directive Version no. 10 (final* version), 15 October 2008, Drafting Group Chemical Monitoring SW, Eur- Comm., DG ENV; http://circa. europa.eu/Public/irc/env/wfd/library?l = /framework_directive/chemical_ monitoring, accessed 31/03/2010.
4. Directive 2008/105/EC of the European Parliament and of the Council of 16 December 2008 on environmental quality standards in the field of water policy, amending and subsequently repealing Council Directives 82/176/ EEC, 83/513/EEC, 84/156/EEC, 84/491/EEC, 86/280/EEC and amending Directive 2000/60/EC of the European Parliament and of the Council., *Off. J. Eur. Union*, **L348**, 24.12.2008, 84.
5. Directive 76/464/EEC of 4 May 1976 on pollution caused by certain dangerous substances discharged into the aquatic environment of the Community, *Off. J. Eur. Comm.*, **L129**, 18.05.1976, 23.
6. Directive 2006/11/EC of 15 February 2006 on pollution caused by certain dangerous substances discharged into the aquatic environment of the Community (codified version), *Off. J. Eur. Comm.*, **L64**, 04.03.2006, 52.
7. Directive 2009/90/EC of 31 July 2009 laying down, pursuant to Directive 2000/60/EC of the European Parliament and of the Council, technical specifications for chemical analysis and monitoring of water status, *Off. J. Eur. Comm.*, **L201**, 01.08.2009, 36–38.
8. S. J. de Mora, *Tributyltin: A Case Study of an Environmental Contaminant*, Cambridge University Press, Cambridge UK, 1996.
9. *A Manual for the Analysis of Butyltins in Environmental Samples*, Virginia Institute of Marine Science, USA, 1996.
10. H. Shioja, S. Tsunoi, H. Harino and M. Tanaka, *J. Chromatogr. A.*, 2004, **1048**, 81.
11. S. Bayen, J. P. Obbard and G. O. Thomas, *Environ. Int.*, 2006, **32**, 915.
12. F. J. Santos, J. Parera and M. T. Galceran, *Anal. Bioanal. Chem.*, 2006, **386**, 837.
13. Z. Zencak and M. Oehme, *Trends Anal. Chem.*, 2006, **25**, 310.
14. E. Eljarrat and D. Barceló, *Trends Anal. Chem.*, 2006, **25**, 421.
15. F. Pellizzato, M. Ricci, A. Held and H. Emons, *J. Environ. Monit.*, 2007, **9**, 924.
16. *European Communities Environmental Objectives (Surface Waters) Regulations 2009*, S.I. No. 272, Department of Environment, Heritage and Local Government, Ireland.

17. S. Sturm, J. Kiefer and E. Eichhorn, in *Veröffentlichungen aus dem Techno-logiezentrum Wasser* [in German], ISSN 1434-5765, 2007, **31**, 1–185.
18. J. Trasher, P. Morgan and S. R. Buss, *Attenuation of Mecoprop in the Subsurface*, Science Group Report NC/03/12; UK Environment Agency, 2004, 100.
19. J. Schuette, *Environmental Fate of Glyphosate. Environmental Monitoring and Pest Management Department of Pesticide Regulation*, Sacramento, CA, USA, 1998, pp. 13.
20. B. Tienpont, F. David, E. Dewulf and P. Sandra, *Chromatographia*, 2005, **61**, 365.
21. O. S. Fatoki and A. Noma, *S. Afr. J. Chem.*, 2001, **54**, 69.

Managing the Water Footprint of Irrigated Food Production in England and Wales

TIM HESS,* JERRY KNOX, MELVYN KAY AND
KEITH WEATHERHEAD

ABSTRACT

This chapter discusses the concept of a water footprint in relation to
irrigated food production in England and Wales and the opportunities to
reduce its environmental impact. It is split into three parts. The first
considers the definition of a water footprint and how it differs from a
carbon footprint. The distinction between different types of water ("blue"
and "green") commonly referred to in water footprint studies and the
problems associated with interpreting the data are described. The second
part reviews the current state and underlying trends in irrigated agriculture, including the volumes abstracted and water sources used. An
important component of a water footprint is the environmental impact of
the water abstracted. Using a geographical information system (GIS),
data on the spatial distribution of irrigation abstractions is combined with
information on water resource availability to identify catchments where
irrigated production (footprint) is likely to be having an environmental
impact. The final part assesses the options for managing water better in
irrigated agriculture, both on-farm and from a catchment (water regulation) perspective. Measures to reduce the water footprint on-farm
include better management (scheduling) and the adoption of new
technologies to improve irrigation application uniformity and water

*Corresponding author.

Issues in Environmental Science and Technology, 31
Sustainable Water
Edited by R.E. Hester and R.M. Harrison
© Royal Society of Chemistry 2011
Published by the Royal Society of Chemistry, www.rsc.org

efficiency. Water regulation measures include limiting abstractions in catchments where environmental damage is known to be occurring, using water trading to reallocate water to higher value agricultural uses and changing the timing of abstraction to encourage winter reservoir storage. Finally, there is a discussion of the implications of increasing our dependence on overseas markets for food supply and the consequences for the agricultural sector's broader role in environmental sustainability and rural employment.

1. Water Footprints – Understanding the Terminology

1.1 Definition of Water Footprint

In the 1990s, a UK researcher popularised the notion that in addition to the physical water content, all agricultural produce and commodities can be assigned a quantity of virtual, or embedded, water that represents all the water consumed in its production.[1] This concept was developed by others[2] who coined the term "water footprint" as a measure of a nation's appropriation of global water resources. It can be considered to be the sum of the virtual water content of the goods and services consumed and is analogous to the concept of "ecological footprint".[3] The water footprint is, therefore, the total volume of freshwater that is used to produce the goods and services consumed by the individual, business or nation.[4] As such, it is a bulk measure of the appropriation of global freshwater resources but it does not indicate the source of the water or the environmental impact. The water footprint of a single product may be derived from a combination of physically (*e.g.* rainfall, groundwater, surface water) and geographically diverse water sources.

The water footprint of a food crop includes all the water used to grow the crop, the water used in post-harvest processing and the water used to assimilate polluting emissions. As plants require large volumes of water to meet the demand for transpiration, the water footprint of food crops can be very large indeed. Estimates for the total water footprint of food crop production in the UK vary. Researchers[5] have estimated the water footprint of the UK's food and fibre consumption. Taking their figures for UK food crop production, suggests a water footprint of $15.1\,\mathrm{Gm^3\,y^{-1}}$ for home grown crops, whilst others[6] estimate the total water footprint of UK agriculture at $9.2\,\mathrm{Gm^3\,y^{-1}}$ out of a total national water footprint of $65.8\,\mathrm{Gm^3\,y^{-1}}$. Agriculture is therefore the largest consumer of freshwater (including rainfall) in the UK and represents 400–700 litres per person per day, far outweighing household consumption at 150 litres per person per day.

1.2 "Blue" and "Green" Water

Such figures are useful to convey the magnitude of an activity's dependence on freshwater systems. However, they tell us nothing of the impact of that water use on global ecosystems. Water used to grow crops can come from a range of sources, and the impact on the environment, or society, will be different for different water sources. In order to make the water footprint estimate more useful, it is common to differentiate between "blue" water, *i.e.* water abstracted from renewable water resources such as rivers, lakes and groundwater, and "green" water, *i.e.* rainfall that is used by the crop at the place where it falls.[7] Most UK crop production is rainfed; therefore most of the water footprint of UK cropping comprises "green" water with a low opportunity cost – if that water were not being used to grow rainfed crops, it would not be available for other uses. Assuming the field is not kept bare, some other vegetation (*e.g.* "natural" vegetation) would potentially use a similar amount of water. There is, therefore, little benefit to be gained by reducing the "green" water component of the water footprint.

Water used for irrigation and post-harvest processing, on the other hand, is "blue" water and has competing uses. It has a higher opportunity cost to society in that if that water were not being abstracted for crop production, it would be available for others to abstract (*e.g.* domestic water supply or industry) or for environmental uses (*e.g.* maintenance of river flows and wet-lands, protected habitats). Even in a humid climate such as in England and Wales, where irrigation is supplemental to rainfall, rising demand for water and increasing competition between sectors is highlighting the threats to "blue" water for agriculture.

2 Water Use in Irrigated Agriculture

In 2005, a baseline assessment of agricultural water use in England and Wales was conducted,[8] and total on-farm water abstraction was estimated to be in excess of $300 \, \mathrm{Mm^3 \, y^{-1}}$, approximately 60% of which was used for irrigation of outdoor field-scale agricultural and horticultural crops ($128 \, \mathrm{Mm^3 \, y^{-1}}$, notably potatoes and field vegetables) or protected and nursery cropping ($53 \, \mathrm{Mm^3 \, y^{-1}}$). Compared with total national abstractions, including those for public water supply and industry, agricultural irrigation constitutes only a minor use (1 to 2%). However, the main problem is that irrigation is a consumptive use (that is, water is not returned to the environment in the short term) and is concentrated in the driest areas in the driest years and driest months when resources are most constrained.[9]

2.1 Areas Irrigated and Volumes of Water Abstracted

Most agricultural cropping in England and Wales is rainfed and, even in a dry year, only a small proportion of land (<5%) is typically irrigated. Since 1955, the government has published statistics on agricultural irrigation in England and Wales, based on surveys carried out roughly triennially, the most of which was performed in 2005.[10] These provide statistics on the areas irrigated, volumes applied and water sources used for irrigation. Data from the last six surveys (1987–2005) are summarised in Figure 1.

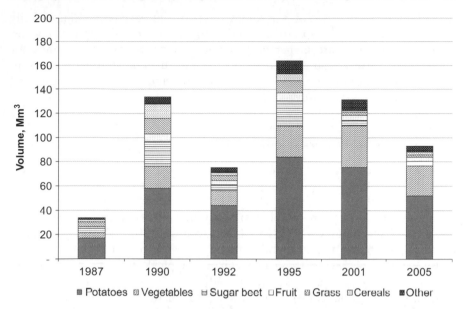

Figure 1   Volumes of irrigation water applied (Mm^3), by crop category, 1987–2005 (Source: Cranfield University, UK).[10]

Over the last 20 years, there have been significant changes in the types of crops irrigated. The proportion of irrigation on grass, sugar beet and cereals has declined steadily. In contrast, there has been a marked increase in irrigation of high value crops, particularly potatoes and field vegetables. This trend is driven by supermarket demands for quality, consistency and continuity of supply, which can only be guaranteed by irrigation. For 2005, the data show that the irrigated areas and volumes of water applied have fallen for almost all crop categories compared to 2001. Notably, the area and volume for irrigated main crop potatoes have fallen by over a third. This partly reflects the recent decline in the total area of potatoes grown in England, which has fallen by almost 19% between 2001 and 2005. Nevertheless, combining main crop and early varieties, potatoes continue to be the dominant irrigated crop, accounting for 43% of the total irrigated area and 56% of water use. Irrigated vegetables have increased slowly in relative share to 28% of the area and 27% of water use. Cereals show an increase in irrigated area, but much less change in water use.[10]

In identifying trends, the data must be interpreted in relation to the weather in each year as summer rainfall patterns and rates of evapotranspiration vary greatly, and hence directly influence the areas irrigated and volumes applied. Using data from 1982 to 2005 the underlying growth rates in the areas irrigated, volumes and depths applied for various crops over time (after allowing for annual weather variation) were analysed[10] using multiple regression techniques. Assuming linear growth, the analysis showed that the total irrigated area of vegetables has been growing strongly at 3% per annum. In contrast, orchard fruit has been declining and soft fruit has remained relatively stable. However, the volumes of water used on vegetables and soft fruit have been rising steadily (3.9% and 2.6% y^{-1}, respectively), reflecting the increased depths of water being applied to ensure higher crop quality. These increasing demands for "blue" water abstraction are placing greater pressures on summer water supplies in some parts of England, notably eastern and south eastern England where irrigation is concentrated.[11]

2.2   Irrigation Water Sources

Most irrigation water is abstracted from rivers and streams, and is used direct with relatively little on-farm storage. It has been reported[10] that half (54%) of all irrigation abstraction in 2005 was from surface sources (rivers, streams). Groundwater abstraction (boreholes) accounted for 41% and other sources (*e.g.* water harvesting) was 4%. Only 1% of water used for irrigation was taken directly from public mains supplies, mainly because peak demands for agricultural irrigation exceed those that can be supplied from the public mains supply. Public mains water is therefore only used on small horticultural units (*e.g.* strawberry production) where lower rates of water use are required, or where water quality is an important factor (*e.g.* glasshouse production). In field-scale irrigation, very little rainwater is harvested and re-used due to the small volumes that can be captured and the requirement for storage, but it is widely used and economically justifiable for small-scale horticultural holdings (*e.g.* protected cropping).

2.3   Location of Irrigation

Irrigation of food crops can be the largest abstractor in some catchments in dry summers and concerns have been raised over the potential impacts of irrigation water abstraction on the environment, particularly in catchments where irrigation abstractions are concentrated and where water resources are under pressure. In many catchments, summer water resources are already over-committed and additional summer licences for surface and groundwater irrigation abstraction are unobtainable.[12]

Information on the spatial distribution of agricultural and horticultural holdings across England and Wales are collected annually by the agricultural levy board (Agriculture and Horticulture Development Board, AHDB) as part of their statutory duty. The type of information and level of detail that is available in the public domain depends on its commercial sensitivity, but the baseline data can be used to map the spatial distribution of growers. In the research presented in this chapter, those involved in the production of the three most important irrigated crop categories have been considered, namely potatoes, field vegetables and soft fruit. Collectively, these account for 85% of the total volume of irrigation water abstracted annually. In England and Wales, the water regulatory authority, the Environment Agency (EA), has assessed the availability of water resources for abstraction at a catchment level. Each catchment has been defined according to its resource status and allocated to one of four categories, "water available", "no water available", "over-licensed" and "over-abstracted", in order of increasing water stress.[12] The spatial distribution of agricultural holdings involved in potato, field vegetable and soft fruit production in 2008 has been mapped and compared with water resource availability, by catchment using a geographical information system (GIS) as shown in Figure 2. The aggregated data by crop sector and water resource category are summarised in Figure 3.

The analysis shows that on average only 10–15% of agricultural holdings are located in catchments where additional water abstraction would be available during summer low-flow periods ("water available"). About half of all holdings are located in catchments defined as either having "no (more) water available" or as being already "over- licensed". Nearly a fifth of all holdings are in catchments defined as being "over- abstracted". Therefore, in water stressed catchments, where water demand for irrigation (abstractions) exceeds available surface or groundwater water supplies, reducing the "blue" water component of the water footprint would mean that water resources could be released to sustain environmental flows or support other uses.

3.   Managing the Water Footprint

In the context of balancing the water needs for agriculture with that of the environment, it is useful to consider the opportunities available to individual growers, the agri-food industry and regulators to reduce the "blue" water footprint. Where abstraction for irrigation of food crops is causing adverse impacts on the water environment or other water users in a catchment, two alternative approaches might need to be considered: either action by a grower to reduce their water use through better water management, or action by the regulatory authority to control or limit irrigation abstraction, or a combination of the two. The measures available and their impact on the water footprint are summarised below.

3.1   Managing Water Better

Internationally, irrigation has a reputation for low water efficiency.[13] However, in the UK, irrigation is supplemental to rainfall. Growers use relatively little water by international standards and are generally highly conscious of the need to improve water efficiency. But even in a humid climate there is scope for using less water in agricultural food production. Making the maximum use of soil moisture and rainfall, knowing precisely where and when irrigation needs to be applied and then applying it accurately and uniformly, are fundamental steps in the "pathway to efficiency" – a process whereby farmers are encouraged to evaluate their performance and move towards best practice.[14] Introducing new technologies and management practices, often developed in more arid countries, together with efforts to bring the average irrigator nearer to the best, can

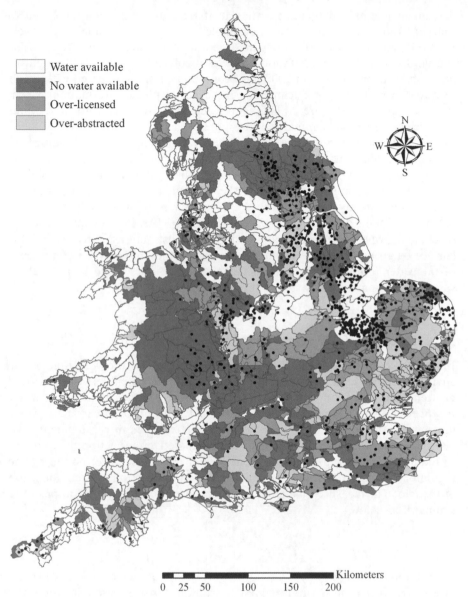

Figure 2 Spatial distribution of agricultural and horticultural holdings involved in potato, field vegetable and soft fruit production and water resource availability, by catchment in England and Wales, in 2008 (Source: Environment Agency, Bristol).[12] Catchments classified as "not assessed" are considered not suitable for the application of the water resource availability methodology developed by the Environment Agency.

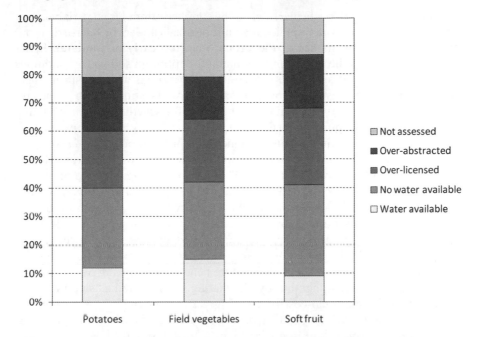

Figure 3   Proportion (%) of agricultural holdings located in catchments with defined levels of water resource availability. Data for England and Wales in 2008.

thus help improve both levels of on-farm water management and reduce the environmental impact of irrigation.

Managing water better predominantly focuses on using water more wisely. Most irrigators already aim to do this because saving water usually means saving money, by reducing pumping, storage and labour costs. But more can be done on-farm to make better use of existing sources, to obtain more "crop per drop" ($t\,m^{-3}$) and more "value per drop" ($£\,m^{-3}$). In England and Wales it is a requirement for renewing agricultural irrigation licences (withdrawal permits) that abstractors demonstrate "efficient" water use.[12] This is proving a strong regulatory measure for driving agricultural water efficiency upwards. However, the cost of pumping water and providing labour and equipment for irrigation also provides a strong economic case for growers to irrigate more efficiently. For example, saving just one application can be financially highly beneficial. Conversely, the costs and implications of inefficient irrigation – poor quality, unsaleable produce – are very significant.[15] The three areas where growers can make better use of water, namely through improved scheduling, better equipment and appropriate use of water resources, are summarised below.

3.1.1   Improving Management to Increase Irrigation Efficiency

Efficient irrigation means optimising soil water management practices to improve crop yield and quality with the appropriate amount and timing of water application. Without proper scheduling it is difficult to keep track of what is happening in the field, particularly when the summer rainfall is so variable. As a result, the amount of water applied may not be ideal (under- or over-irrigation) or the water may be applied at the wrong time (too early or too late). Better scheduling provides the opportunity to maintain optimum soil water conditions for crop growth, in order to meet yield and quality targets whilst ensuring there is minimum water wastage. However, many growers still rely on subjective methods (such as visual observation or intuition) for scheduling their irrigation and a significant proportion do not schedule using any form of objective (scientific) technique (such as water balance computer models or *in situ* soil moisture measurement). This is reflected in a recent survey of irrigation practices in England[10] which showed that only 60% of the total irrigated area nationally is scheduled using objective (scientific) approaches. The relatively poor uptake of scheduling remains a reflection of the difficulty of being able to accurately quantify (or convince farmers) of the benefits that it provides, particularly in dry years when many growers are more constrained by water resources and/or equipment rather than by water management decision-making (scheduling). Most growers believe that they will get more benefit from risking over-irrigating rather than under-irrigating; thus over-irrigating is considered a "safe" strategy. However, as the marginal cost of applying water rises (mainly due to energy prices) not only is over-irrigating a waste of water resources with implications for the water footprint, but it also becomes a costly exercise.

In countries such as the UK, rainfall is "free irrigation" but this advantage can easily be lost through poor scheduling. Irrigating to a fixed schedule, irrespective of weather, or more commonly, failure to allow adequate storage capacity in the soil for unforeseen rainfall following irrigation, means that rainfall is not used to its full effect, increasing the volume of water abstracted and potentially causing excess leaching and groundwater pollution.[16] Irrigating to a planned deficit can make better use of rainfall, particularly when used in combination with weather forecasting to avoid rainfall losses, and with appropriate soil tillage practices to improve soil structure for water retention. Many of the emerging in-field water management (scheduling) technologies and control systems were first developed overseas for application in more arid and semi-arid environments (*e.g.* Australia) and then modified for UK cropping conditions.

3.1.2   Switching Technology to Increase Irrigation Application Uniformity

Good equipment is essential to apply water uniformly and adequately. Modern systems such as booms, close-spaced solid-set sprinklers and drip (or trickle) irrigation can potentially improve water application efficiency by applying water more uniformly and minimising losses (evaporation, deep drainage, surface runoff). Achieving higher levels of uniformity enables water applications to be scheduled more accurately, which in turn leads to water saving, reduced costs and better quality crops.

It is widely acknowledged that achieving greater efficiency of water use is an important aspect of water resource management at the catchment scale, but caution needs to be exercised in promoting one method of irrigation over another at the farm level. For example, although drip irrigation is *potentially* more efficient than overhead methods (*e.g.* sprinklers), in practice its *actual* efficiency is often much lower than expected, particularly where levels of on-farm water management are inadequate or inappropriate.[17] Similar problems have also been widely reported internationally. With a permanent automated system, such as trickle irrigation applying water directly into the soil, it is very easy to leave it running for too long, leading to deep percolation losses. Trickle irrigation users are also usually more focused on maximising crop yield and quality than saving water *per se*.

3.1.3 Securing Water Resources and Using "Appropriate" Quality Water

Although water is becoming scarce in many catchments in England at critical times of year (see Figure 2), there are still opportunities for improving the sustainable use of water resources. Surprisingly, most of the water allocated to irrigated agriculture is not actually used, even in dry years, due to a variety of agronomic, economic and water-resource constraints.[18] In many instances, the water is in the wrong place and/or available at the wrong time. Finding environmentally and financially viable methods of transferring it to where it is needed, or storing it for use in the following summer, would go a long way towards resolving present water shortages. One of the most popular strategies for increasing water security on-farm is through the use of storage reservoirs (either individual or shared), harvesting high flows during the winter and storing for use the following summer. Many growers have switched from summer to winter abstraction to provide greater security and flexibility of supply, enabling them to better balance their water supply needs with changing crop demands. Charges for winter abstraction are also significantly lower, providing a regulatory incentive to switch sources. Storing water therefore acts as an "insurance policy" for the grower as well as potentially providing an additional business income through water trading. Although storage would slightly increase the total water footprint due to leakage and evaporation losses, it should reduce the environmental impact of the water abstraction, reducing pressure on summer supplies when resources are most stressed, in addition to enhancing the local landscape and providing new opportunities for biodiversity.

There are substantial opportunities to obtain better value from available water by trading or sharing resources, by promoting water benchmarking[19] to improve irrigation performance and by developing conjunctive use of surface water and groundwater to improve reliability of supply. Whilst such measures do not necessarily result in reductions in water use they do improve levels of water use efficiency ($t\ m^{-3}$) and water value ($£\ m^{-3}$) within agriculture. Growers need to assess which combination of measures is technically and economically appropriate for their circumstances and then prioritise accordingly.

3.2 Managing Abstraction

Becoming more "efficient" at the farm level may not necessarily reduce the water footprint of irrigated food production if the water that is released is used to increase the irrigated area or the depth applied. In some catchments, it may be necessary to manage abstraction by regulation.

All water abstractions in England and Water greater that $20 \, \text{m}^3 \, \text{d}^{-1}$ require an abstraction licence from the water regulatory authority, the EA. New licences and renewals are set in accordance with the Catchment Abstraction Management Strategy (CAMS) developed for each catchment, for both surface and groundwater. Tiered "hands-off" flow or groundwater level limits protect the sources from over-abstraction. If conditions alter, for example due to climate change, the regulator will be able to revise these limits periodically to take account of the changes. Unfortunately, many of the older licences did not include such constraints and most were not time-limited. In those cases, the EA can still stop abstraction through "Section 57" restrictions and Drought Orders, though these are blunter instruments as all abstractors are treated in the same way. The regulator is currently looking at mechanisms whereby all licences become time-limited, though there is understandably considerable farmer opposition.

In over-licensed or over-abstracted catchments where abstraction is causing a significant environmental impact, the EA are implementing a programme for restoring sustainable abstraction (RSAP), variously by amending licensed volumes, abstraction points or other details. Abstractors currently receive compensation if voluntary charges cannot be agreed; however, after 2012 the licences can be amended or removed without compensation where "serious" environmental damage is being caused.

Together, these arrangements give the regulator sufficient powers to stop major environmental damage occurring due to abstraction and so, in theory at least, the associated water footprint should not be causing serious environmental harm. However, damage due to the combined effect of many abstractions is harder to identify and remove. Furthermore, the arrangements do not yet give the regulator powers to move licences to the most beneficial users, so the footprints may be far from optimal. Promotion of abstraction licence trading (water trading) gives abstractors some scope, but this is aimed more at economic efficiency than environmental efficiency.

4   Discussion

It is important to remember that the volumetric water footprint is a measure of the appropriation of water resources and is not in itself a measure of the impact of that allocation on the environment or society. In unstressed catchments or in locations where there are no competing downstream users (including environmental flows) there may be no benefit from simply reducing the volumetric water footprint and it could, in fact, have an adverse impact on the agri-food sector. It is more appropriate, therefore, to concentrate on strategies that reduce the impact rather than the size of the water footprint.

Despite its small land allocation, irrigated production in England and Wales is high value, accounting for approximately 20% of total crop value and in some parts of the country is a key component of the rural economy.[20] For example, in eastern England more than 1000 agri-businesses depend on water to supply high quality produce to the nation's supermarkets, providing over 30% of potatoes and 25% of all fruit and vegetables. In this region, it is estimated that the agri-food industry employs over 50 000 people (both on-farm and in downstream support industries) and contributes some £3 billion annually to the region's economy. Water is at the heart of this industry and without it many farmers would simply not be able to meet the exacting standards of quality and continuity of supply demanded by supermarkets and consumers.[11] Irrigated agricultural production thus has a critical role to play, both in terms of contributing to national food production, the economy and rural employment.[21,22]

There are also issues surrounding the links between water footprints and food security. At present, the UK has the security of being able to produce approximately half of all the food consumed in the UK.[23] However, the UK has been ranked in the world's top six virtual water importing countries.[24] In one sense this offers security against drought, but it also makes the UK vulnerable to droughts in those countries that supply the UK with fresh food. In the global market, water now connects people across the world in ways that were unimaginable 50 years ago. Importing food from places where water is scarce can be seen as exporting drought and environmental problems to places where there are fewer regulations, poor environmental protection and a lack of financial resources to deal with the problems this can create. It may also cause food shortages in the exporting countries and drive up local food prices. Policies which support home production are important for the UK's food security, but food must be produced in ways that protect and enhance the natural environment.[25]

Finally, it is also important to recognise the role that agriculture plays in the UK economy. Agriculture has a multifunctional role, sitting at the interface between ecosystems and society, contributing to a range of "non-food" services including landscape enhancement, leisure and recreation, as well as "food" production. It is because of these interactions and feedbacks that any assessment of the risks (*e.g.* climate change) and impacts (*e.g.* water footprint) of agriculture on the environment is notoriously difficult. Producing food sustainably in a changing and uncertain climate will be a priority – but dealing with the externalities on agriculture needs to be handled in ways that are sensitive to both ecosystems and the diversity of benefits that agriculture provides, and not just to food production. Recent concerns regarding possible future global food shortages, exacerbated by climate change, have raised questions about food security at a national scale. The UK government does not prescribe targets for national self sufficiency, but seeks to achieve "food security" by guaranteeing household access to affordable, nutritious food.[23] UK agriculture, along with the food industry as a whole, is charged with "ensuring food security through a strong UK agriculture and international trade links with EU and global partners which support developing economies".[26] In this regard, UK agriculture is required to be internationally competitive, whether this is delivering to domestic or international markets. Factors such as climate change and consumer demands for knowledge on water footprints could affect not only the relative productivity of UK agriculture and the demand for water for food production,[27] but also its competitive position in international markets. Caution therefore needs to be exercised in the derivation of water footprint data and its policy implications.

5 Conclusion

Agriculture requires large volumes of freshwater and has the largest water footprint of any sector in England and Wales. Understanding the magnitude of the water footprint of irrigated crop production can help to contribute to sustainable food production by identifying activities that appropriate large volumes of water. However, unlike "carbon footprints", the "water footprint" needs further elaboration before it becomes a useful indicator. It is important to separate "blue" from "green" water, but it is not all about "size" – large water footprints are not necessarily "bad". It is important to relate the magnitude of the water footprint to its impact on society and the environment. It is clear that in some parts of England and Wales, and at some times of year, the abstraction of water for irrigated crop production places considerable stress on freshwater resources. There are opportunities to alleviate this stress, by making better use of water on-farm, or through better abstraction control. However, the contribution of irrigated agriculture to domestic food production and rural livelihoods must not be ignored. Growers, regulators and the food industry need to work together to encourage benign water use, rather than simply reduce the "water footprint" in particular crop sectors.

Acknowledgements

The authors acknowledge the Potato Council (PCL) and Horticultural Development Company (HDC) for provision of grower data, the Environment Agency for data on water resource availability and Dr. Andre Daccache for the GIS analyses.

References

1. J. A. Allan, Virtual water-the water, food and trade nexus: useful concept or misleading metaphor? *Water Int.*, 2003, **28**, 4–11.
2. A. Y. Hoekstra and P. Q. Hung, Virtual water trade: a quantification of virtual water flows between nations in relation to international crop trade, *Value Water Res. Rep. Ser. No. 11*, UNESCO-IHE, Delft, The Netherlands, 2002.
3. M. Wackernagel and W. E. Rees, *Our Ecological Footprint: Reducing Human Impact on the Earth*, The New Catalyst Bioregional Series 9, New Society Publishers, Canada, 1996.

4. A. K. Chapagain and A. Y. Hoekstra, Water footprints of nations, *Value Water Res. Rep. Ser. No 16*, UNESCO-IHE, Delft, the Netherlands, 2004.
5. A. K. Chapagain and S. Orr, *UK Water Footprint: The Impact of the UK's Food and Fibre Consumption on Global Water Resources*, WWF-UK, 2008, vol. 1.
6. Y. Yu, K. Hubacek, K. Feng and D. Guan, Assessing regional and global water footprints for the UK, *Ecol. Econom.*, 2010; Doi: 10.1016/j.ecolecon. s009.12.008.
7. M. Falkenmark, Land-water linkages: a synopsis, in Land and water integration and river basin management, *FAO Land Water Bull.*, FAO, Rome, Italy, 1995, No. I, 15–16.
8. J. King, D. Tiffin, D. Drakes and K. Smith, *Water Use in Agriculture: Establishing a Baseline*. Final Report, Project WU0102, Defra, UK, 2005.
9. J. W. Knox, E. K. Weatherhead, J. A. Rodríguez Díaz and M. G. Kay, Development of a water-use strategy for horticulture in England and Wales – a case study, *J. Horticul. Sci. Biotechnol.*, 2010, **85**(2), 89–93.
10. E. K. Weatherhead, *Survey of Irrigation of Outdoor Crops in 2005: England and Wales*, Cranfield University, Cranfield, UK, 2006.
11. J. W. Knox, E. K. Weatherhead, J. A. Rodríguez Díaz and M. G. Kay, Developing a strategy to improve irrigation efficiency in a temperate climate – A case study in England, *Outlook Agric.*, 2009, **38**(4), 303–309.
12. Environment Agency, *Demand for Water in the 2050s – Briefing Note*, Environment Agency, Bristol, 2008, pp. 12.
13. C. Perry, Efficient irrigation; inefficient communication; flawed recommendations, *Irrigat. Drainage*, 2007, **56**(4), 367–378.
14. J. W. Knox, Understanding irrigation efficiency, *UK Irrigat.*, 2004, **32**, 4–17.
15. J. W. Knox, J. Morris, E. K. Weatherhead and A. P. Turner, Mapping the financial benefits of sprinkler irrigation and potential financial impact of restrictions on abstraction: a case study in Anglian Region, *J. Environ. Managem.*, 2000, **58**, 45–59.
16. T. M. Hess, Minimising the environmental impacts of irrigation by good scheduling, *Irrigat. News*, 1999, **28**, 3–10.
17. J. W. Knox and E. K. Weatherhead, The growth of trickle irrigation in England and Wales; data, regulation and water resource impacts, *Irrigat. Drainage*, 2005, **54**, 135–143.
18. E. K. Weatherhead, J. W. Knox, J. Morris, T. M. Hess, R. I. Bradley and C. L. Sanders, *Irrigation Demand and On-Farm Water Conservation in England and Wales*, Final Report to MAFF, Cranfield University, Cranfield, UK, 1997.
19. J. A. Rodríguez Diaz, E. Camacho Poyato and R. Lopez Luque, Applying benchmarking and data envelopment analysis (DEA) techniques to irrigation districts in Spain, *Irrigat. Drainage*, 2004, **53**(2), 135–143.
20. J. Morris, K. Vasileiou, E. K. Weatherhead, J. W. Knox, and F. Leiva-Baron, The sustainability of irrigation in England and the impact of water pricing and regulation policy options, Proceedings of International Conference on Advances in Water Supply Management, Imperial

College, London, UK, 15–17 September, *Adv. Water Supply Managem.*, 2003, 623–632.

21. W. Leathes, J. W. Knox, M. G. Kay, P. Trawick and J. A. Rodríguez-Diaz, Developing UK farmers' institutional capacity to defend their water rights and effectively manage limited water resources, *Irrigat. Drainage*, 2008, **57**, 331–332.

22. A. Angus, P. J. Burgess, J. Morris and J. Lingard, Agriculture and land use: demand for and supply of agricultural commodities, characteristics of farming and food industries and implications for land use, *Land Use Policy*, 2009; doi 10.1016/j.landusepol.2009.09.020.

23. Defra, *UK Food Security Assessment: Detailed Analysis,* HMSO, London, 2010.

24. A. Hoekstra and A. K. Chapagain, *Globalisation of Water Sharing the Planet's Fresh Water Resources,* Blackwell, UK, 2008.

25. H M Government, *Food 2030. How We Get There,* H M Government, January 2010.

26. Defra, *Agriculture in the UK,* HMSO, London, 2009, pp. 146.

27. Environment Agency, *Water Resources in England and Wales – Current State and Future Pressures,* Environment Agency, Bristol, 2008, pp. 23.

Social Justice and Water

ADRIAN MCDONALD,* MARTIN CLARKE, PETER BODEN
AND DAVID KAY

ABSTRACT

Social justice and water refers to fairness of access to water resources and equality of burden from poor water quality and water hazards. This chapter considers briefly the origins and definition of social justice, and ways in which social justice and water interact. Global scale water justice is considered mainly in respect of disease and the lack of progress towards effective provision of water and sanitation. A detailed case study of the links between water debt and deprivation in the UK is then examined, concluding with a consideration of the options available to resolve water debt issues whilst respecting social justice. We conclude with a consideration of the future constraints on water resources in the light of the latest UK climate projections, and recognise that difficult choices might be encountered in which social justice for people might have to have priority over environmental justice for river and land habitats.

1 The Emergence of the Social Justice Concept

Social justice as an unattributed concept has a history of several hundred years but its codification is argued by some to lie in the Catholic Church in the middle of the last century.[1] Social justice does not come alone. It is part of a larger agenda of change towards a more egalitarian society and as such is a political construct. There is a hierarchy founded on the general concept of social justice and becoming more specific as it passes first through environmental justice and finally to resource- or issue-specific matters such as, in this instance, water

---

*Corresponding author.

Issues in Environmental Science and Technology, 31
Sustainable Water
Edited by R.E. Hester and R.M. Harrison
© Royal Society of Chemistry 2011
Published by the Royal Society of Chemistry, www.rsc.org

justice. Paralleling each of these stages is the other side of the coin: social, environmental and water injustice.

In reality, communities, companies and countries do not often fall into one camp or one extreme but will lie at a compromise point in the spectrum between justice and injustice, between equality of access and ability to pay, between burden bearing and ability to escape or acquire protection, either physical, such as flood proofing, or economic, such as insurance.

2 Definitions and Meaning

There are many definitions of environmental justice. Some tend to be all encompassing, covering all attributes of the environment where we work, live and visit, and it appears to be a concept *weakened* by that broadness. At its heart environmental justice is about fairness. That fairness can relate to equality of access (to positive benefits), equality of protection (from negative impacts) and equality of influence (in decision making). Social justice refers to "the distribution of advantages and disadvantages within society".[2] The "default (Wikipedia!) definition" of social justice is more specific:

> *"A world which affords individuals and groups fair treatment and an impartial share of the benefits of society."*

Environmental justice is yet more specific. Environmental justice refers to the equitable distribution of environmental burdens and benefits. However environmental justice activists are usually concerned with the opposite perspective: injustice. This lack of justice is claimed to be borne, primarily, by identifiable groups such as (i) racial minorities, (ii) women, (iii) residents of economically disadvantaged areas, or (iv) residents of developing nations.

The US Department of Transportation[3] offers the following definition of the three environmental justice principles:

1. To avoid, minimise or mitigate disproportionately high and adverse human health and environmental effects, including social and economic effects, on minority populations and low-income populations.
2. To ensure the full and fair participation by all potentially affected communities in the transportation decision-making process.
3. To prevent the denial of, reduction in, or significant delay in the receipt of benefits by minority and low-income populations.

3 Water and Interaction with People

Water interacts in several ways with people:

1. Water is a vital resource for life and for agriculture.
2. Water is an important component of effective sanitation.
3. Water is a threat to well-being when it floods.

4. Water that is contaminated is a risk to health.
5. Sight and sound of water adds to well-being and to house value.

People may be excluded from this resource by drought or by price. They will be subject to the hazards associated with water by their knowledge and ability or otherwise to move rapidly or relocate to a safer site.

4 Water and Social Justice on a World Scale

At least a billion people had inadequate access to water and sanitation in 1980 when the International Water Supply and Sanitation Decade was launched.[4] This is a figure that has remained stubbornly at this value for several decades. There have been some improvements in some regions and supply has advanced more rapidly than sanitation, but water-related problems continue. Figures today would suggest at least 1.1 billion without access to good water and over 2 billion without access to adequate sanitation. Over 2 million people die each year from water-related diseases and a much bigger loss of meaningful productive healthy years – perhaps 80 million work years are lost each year.[5]

The improvements in water supply being more marked than those in sanitation do not fit well with a seminal study in 1996 (ref. 6), which examined the health impact of improved sanitation and improved water provision in a sample of *ca.* 12 000 rural and *ca.* 5000 urban dwellers in eight developing countries in Africa, Latin America and Asia. The study concluded that improvements in sanitation had a significant impact on the health of the subjects, but that improvements in water supply was less influential and would be masked by lack of sanitation improvements. In essence, the paper both confirmed the link between poor water and poor health in developing countries and set a path for prioritisation of intervention that, in essence, was sanitation first and supply improvements second. The sanitation influence appears to relate to the enhanced spread of communicable diseases by rainfall episodes in the absence of adequate drainage.[7] These significant conclusions were reinforced and clarified by a more recent systematic review and meta-analysis, sponsored by the World Bank and comprising 2120 publications and 46 sentinel studies.[8]

Possibly the most stark, yet most encouraging, example of water-related injustice in developing countries relates to the Guinea worm. In the 1980s, the Guinea worm infected millions of people each year. The Guinea worm is ingested within a small water flea or arthropod. Initially in the digestive tract, the female Guinea worm (a parasitic nematode *Dracunculus medinensis*) emerges by burrowing downwards and emerges usually in a lower limb. It is a non-fatal, cripplingly painful infection known as the "fiery serpent". It is not an infection to which immunity is developed and people can be repeatedly infected – those same poor people who have little choice but to use infected stagnant water. Secondary infections of the exit wound can, however, be fatal. But the positive news is that it can be controlled by education-induced behavioural changes – filtering the drinking water through fine gauze, using safer sources such as ground waters, and banning people with emergent worms from using and contaminating the water source. Through these simple community approaches, infection numbers dropped to fewer than 5000 cases in 2008, mainly in southern Sudan.[9]

5 Flooding and Social Justice

Flooding is impartial in terms of who it impacts upon, being driven by physical processes having no regard for people and property. Yet we see major differences in the effects of inundation and on the provision of defences erected against floods. Floods of a similar magnitude and similar warning times can result in very different impact characteristics. In the US, such floods have major economic impacts arising from insurance claims and disrupted production but tend to involve little loss of life. In contrast, floods in the third world have much lower economic impact but have much greater loss of life (see Table 1).

Such a generalised statement, however, can hide exceptions. In August 2005, Hurricane Katrina caused major flooding to the city of New Orleans and, in response, the US authorities appeared muted. It could, of course, simply have been that the authorities were overwhelmed by the scale of the disaster resulting from the extreme severity of the event. But there arose an immediate under-current that suggested that the slow and inadequate response might also be related to the nature of the city involved, which was poor, southern and predominantly black. This may or may not have been the case, but perceived injustice can be as corrosive as real injustice. The social injustice associated with Katrina has been widely examined[10] and the general conclusion appears to be that although Katrina did not create inequalities it highlighted those that exist.

In the UK, flood defence funding is allocated through a scoring system. The scoring system has been recently updated by Defra (Department of Environment, Food and Rural Affairs) but at its heart it remains sensitive to populations involved and so population density is important in arriving at a score. The result is that flood defence is more likely to be funded in urban situations than in rural areas. Rural areas, despite perceptions to the contrary, are not

Table 1 Flood impacts in the First and Third World.

Metric	First World	Third World
Loss of life	Typically low. Very seldom exceeding 100.	Typically high. Losses in thousands.
Financial Impacts	Substantial. For major events – billions of pounds.	Modest. Often not calculated.
Support	Emergency services within hours. Insurance and government intervention.	Limited emergency support based on self or community help. Little government or insurance support.
Evacuation	Common. Automated warnings, individually targeted although seldom comprehensive.	Evacuation options limited. Warnings have limited distribution.

uniformly affluent and rural poverty and lack of access to facilities can be a major issue. In both rural and urban contexts, working class housing tends to be on the flat land in proximity to the factories and prone to floods, whereas the better housing tends to be higher (originally located away from the factory smoke) and so is less likely to be flooded. Examples might be the floods in Carlisle, Cumbria and Hawick, Scottish Borders, although like so many issues in the complex, diverse world of water resources there are exceptions to any rule and in the last decade we have seen a proliferation of new, up-market riverside developments which start to modify the existing social distributions and flood impacts. Thus we see the compromise between water location benefits and impacts.[11]

In relation to floods, we may conclude that the impact of floods and the allocation of flood defence resources may not be uniform, but in the UK at least there is little evidence that there is a current systematic social bias, but there may be some residual, historic, unintended bias which is not spatially homogeneous.

6   Water and Social Justice on a UK Scale

6.1   What Price Water?

Table 2 lists the average bills for water across the UK water companies (Water UK, *personal communication*). There is considerable price variation driven not by differences in water usage but by tariffs. South West Water has historically had the largest bills. These bills are large for at least two reasons: the first, because the region has an extensive coastline and so produces disproportionately high environmental bills to ensure bathing water's compliance. Secondly, the population using water services is highly variable, with a marked summer peak due to tourism. This peak requires installed capacity to be able to meet this peak demand, a financial investment that is not required throughout

Table 2   Average annual bills in 2008 for Water and Sewage Companies (WASC) and Water Only Companies (WOC) in England and Wales.

WASC	Average bill (£)	WOC	Average bill (£)
Anglian Water	361	Bournemouth and West Hampshire	134
Dwr Cymru	374	Bristol Water	168
Northumbrian Water	331	Cambridge Water	116
Severn Trent	291	Cholderton & District	188
South West Water	483	Dee Valley Water	130
Southern Water	393	Portsmouth Water	87
Thames Water	313	South East Water	174
United Utilities	364	South Staffordshire Water	126
Wessex Water	424	Sutton and East Surrey Water	167
Yorkshire Water	332	Veolia Central	146

(Source: Water UK, *personal communication*).

OVER THE YEARS THE BALANCE HAS SWUNG

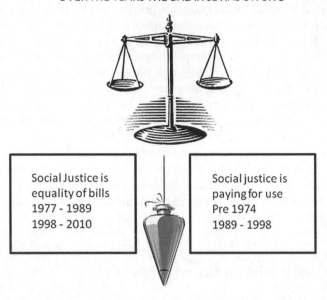

| Social Justice is equality of bills 1977 - 1989 1998 - 2010 | | Social justice is paying for use Pre 1974 1989 - 1998 |

Figure 1   Swinging between interpretations of social justice.

the closed season. So the customers of the south west, a large proportion of whom are retired, are effectively subsidising good environmental quality on behalf of millions of visitors from other UK regions and overseas.

The debate surrounding water justice and price equalisation across regions is not new. Over the last 40 years the pendulum has swung in the UK between a benefits principle and an equity principle as illustrated in Figure 1.

Before the creation in 1974 of the Regional Water Authorities in the UK, prices varied between water suppliers but these were effectively masked: (i) by being aggregated with local council rates bills and (ii) by the sheer numbers of utilities involved (well over 1000 water supply and sewage entities). Post 1974, price differences between the now-small numbers of Regional Water Authorities was evident to consumers and politicians and raised concern. It was proposed that there should be an equalisation of bills between the different regions, implying a significant cross-subsidisation between regions. This was put into effect by the government which initiated first a review of water charges[12] and subsequently the passing in 1977 of a Water Charges Equalisation bill, designed to put into effect the direct subsidisation between regions advocated by the review. In effect, under the terms of the Act, those Water Authorities with the lowest costs (*i.e.* all with a below-average cost) were required to pay a levy to the National Water Council which was to be transferred directly to companies with costs above average. This was put into effect during 1978–81 and amounted to some 2% on unmetered bills – a small but not insignificant subsidy.[13] This was never a recipe for an efficient water sector, because there was limited review of the causes of the differences in the cost base which are a mix of local circumstances and differences in management efficiency. This cross-subsidy was seen as an unfair action and, under the next Conservative administration, the Water Authorities were split up to create a regulator (the precursor of the present Environment Agency in England and Wales) and a series of privatised water companies with the freedom to charge different prices. The following Labour administration did not change this but there is on-going pressure towards price equalisation.

So the question arises as to what social justice in respect of water actually is? Is it paying for the water you take from the environment, with weak parallels to the "Polluter Pays Principle"?[14] Or is it equality of price for a unit of the same product? Water utilities are often gathered with the gas and electricity industries for purposes of comparing approaches and performance, and these industries have the same price across the UK for the same company, but as several companies operate across the UK and have different tariff structures there is a choice and, for some people, choice and freedom to choose are important components of social justice.

Water is a different and, in many ways, unique product and industry in the UK for several reasons. Firstly, water cannot be bought from a local shop in the volumes required. Secondly, it is a monopoly supply to domestic customers, unlike other utilities such as gas and electricity. Thirdly, it is vital for the sustenance of life as without water, even for a relatively short period, we die, and fourthly, with limited quantity or quality of supplies there are potential health consequences. So the industry has clear responsibilities.

6.2   Water and Social Justice in the UK at Fine Scale

6.2.1   Why an Analysis of Water Debt: Water Debt and Corporate Justice?
This section considers an analysis of water and water debt carried out by the authors but here we focus on the social justice ramifications. The original study for United Kingdom Water Industry Research (UKWIR; ref. 15) had explicit corporate equity considerations as drivers for the investigation. The UK regulators in general and Ofwat in particular require companies to deliver a set of metrics on their performance. Ofwat (the Water Services Regulation Authority) is the economic regulator of the water and sewerage sectors in England and Wales. It exists to ensure that the companies provide household and business customers with a good quality service and value for money, and employ comparative metrics to help them achieve this end. These metrics are supplied both annually as June returns and quinquennially as part of the Asset Management Planning exercise, an important exercise that determines water pricing for the subsequent five years. The regulator frequently simply tabulates the performances and then queries why some companies have particularly high or low metrics (often deemed as poor performance) and requires companies to invest further in particular issues. One metric relates to water debt and Ofwat has criticised some companies reporting larger outstanding debts than others. The companies saw this as poor analysis and

unjust because it did not consider the debt performance in the context of the differences in level of deprivation in the company areas. In effect the companies sensed a corporate injustice.

6.2.2 Water Debt in Context

Water debt in the UK is approximately triple the size of the debt held by other utilities, such as gas and electricity, despite water bills being about one third of energy bills, but in relation to overall household debts the water debt is modest. Table 3 (ref. 15) provides information on the changes in household debt in recent years and the relative importance of water debt to overall debt.

The debt situation is set to worsen considerably as the recession and the period of lowered economic activity continues and may be evidenced as follows:

1. 500 000 households were expected to be more than 3 months in arrears on mortgage payments by end of 2009.
2. 11% of borrowers missed payments on their mortgage, credit card or personal loan in the last six months of 2008.
3. 12% of consumers missed a household bill (gas, electricity, water, council tax, telephone, rent) in the last six months of 2008.

Figure 2 shows the growth in the revenue outstanding to the water industry between 2004 and 2008 (ref. 15). The debt nearly doubled in this four-year period and currently this represents an average debt of £53 per household. However, the last decade has, for the most part, been a period of marked expansion and a substantial building programme has been undertaken that has seen extensive development of city centre flats as well as out-of-town new developments. So this debt increase may be partly explained by increases in the number of properties. However, as Figure 3 demonstrates, water industry debt has increased faster than the rate of new property development.[15] Figure 3 indexes property numbers and debt levels to 100 in 2004 and explores the changes in the subsequent years. Note that throughout the text we have used the terms "metered" and "unmetered". We have retained in Figures 3, 8 and 9

Table 3   Household and water debt characteristics 2005 to 2008 (ref. 15)

	2005	*2008*	*Increase %*
UK Personal Debt	£1 158 000 000 000	£1 457 000 000 000	26
Mortgages	£965 200 000 000	£1 224 000 000 000	27
Unsecured Lending	£192 800 000 000	£233 000 000 000	21
Average Household Debt			
Excluding Mortgage	£7786	£9950	23
Including Mortgage	£46 863	£59 700	27
Excluding mortgage but where household has an unsecured loan		£21 766	
Water Debt	£827 000 000	£1 225 000 000	48
% UK Personal Debt	0.07	0.08	

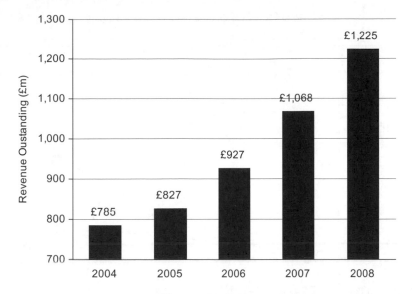

Figure 2 Growth in water industry debt in England and Wales 2004 to 2008 (ref. 15).

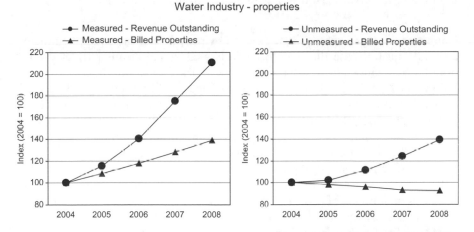

Figure 3 Growth in water debt against growth in water using properties indexed to 100 in 2004 for both metered and unmetered households.[15]

the terms "measured" and "unmeasured" as the water industry sees small, mainly philosophical, differences in this terminology and as this was the definition in the data from which the figures are derived. Whether the information is considered for "metered" or "unmetered" properties, the same picture of debt advancing faster than properties emerges. In fact, the number of unmetered properties has declined as a proportion, yet their associated debt has risen in both absolute and relative terms. This returns us to the "can't pay – won't pay" split that exercised the House of Lords Science and Technology

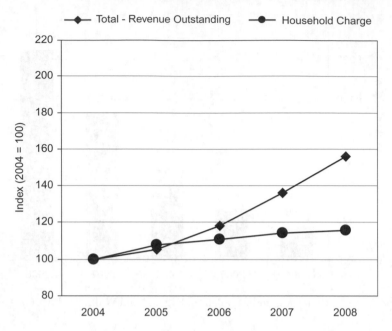

Figure 4   Growth in water debt against water bills indexed to 100 in 2004 (ref. 15).

Committee.[16] If the debt is largely owned by people who can't pay then there is a social injustice case that has to be addressed; a significant part of the UK population cannot afford a basic resource and are only supplied by the legislation which prohibits the cutting off of supplies and by accumulating debt. In contrast, those who won't pay are transferring debt onto others who do pay and to the shareholders who have invested in the company. Clearly a social justice issue remains but is borne by a different population and, if the "won't pay" population are not evenly distributed across companies there is a corporate justice issue in addition.

Perhaps, however, the driver of increased debt is neither inability to pay nor increases in numbers of properties but is simply the increase in the bills. Figure 4 compares the trends in bill sizes with the trends in debt, again indexed to 100 in 2004 (ref. 15). Clearly debt is increasing faster than the bills. In summary then, it appears that debt is increasing in metered and unmetered properties and is largely unrelated to either the numbers of properties or the size of the bills.

6.2.3   The Linkage to Deprivation
To explore further the debt issue and its links to deprivation, and so to social justice, requires a large sample of customers' payment characteristics from the companies and information on deprivation. Figure 6 outlines the weightings used to generate the index of multiple deprivation (IMD). The IMD in England is based upon seven measures of education, income, health, employment, crime, access and environment, each of which is differently weighted as shown in

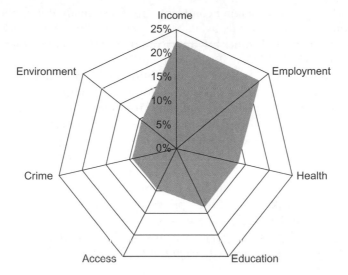

Figure 5 Information sources for water customers and multiple deprivation.[15]

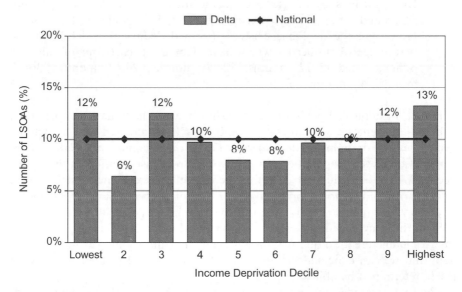

Figure 6 Income deprivation by deciles for company delta region.[15]

Figure 5 (sourced from the Department of Communities and Local Government). Similar indices of deprivation exist for Scotland and Wales but are based on slightly different mixes of measures and weightings and so are not directly comparable. In England, these data are available for Lower Super Output Areas (LSOA) at both an aggregate level and at the individual component level. The LSOA is the smallest spatial unit through which the census results are reported whilst still maintaining confidentiality for individual households. The deprivation index is, therefore, available at relatively fine granularity, with an index for small spatial zones of approximately 1500 people thus yielding approximately 33 000 LSOAs in England and Wales.

For each water company area, the distribution of deprivation in each of 10 equal categories was determined. Figure 6, for a real but anonymous company, shows the real deprivation in that company from the least deprived (lowest) to the most deprived (highest).[15] The line of equal distribution is the national line. Figure 7 provides the same data for every UK water company investigated[15] (covering in excess of 80% of the population). Now we can compare the deprivation profiles and note that some companies (beta and gamma, for example) peak in the most deprived categories whereas others (phi, epsilon, lambda and mu) peak in the least deprived deciles. Some companies have no areas in the lowest deprivation deciles.

To make the link to social justice we now need to examine some metrics of water debt that are associated with each LSOA. We can examine three metrics:

1. How many people owe? [termed "debt penetration", or the percentage of properties which are indebted (>3 months and £10 thresholds)].
2. How much is owed? [termed "debt intensity", or the average debt associated with an indebted property (>3 months and £10 thresholds)].
3. How long the debt has existed? [(termed "continuing debt", or the percentage of properties which have debt in at least four of the time periods listed (3–12 months, 12–24 months, 24–36 months, 36–48 months, 48 months +)].

Figure 8 examines directly the relationship between how many people are in debt ("debt penetration") and deprivation for both metered and unmetered properties.[15] For unmetered properties (70% of England and Wales), debt penetration rises as deprivation levels increase, from just over 3% in the lowest deciles to 18% in the highest: a six-fold increase in the prevalence of debt. In the smaller, metered, cohort, approximately 7–8% of properties are indebted even in areas where income deprivation is lowest, and debt penetration increases to 20% in those areas with the highest income deprivation score a value similar to that for unmetered. So deprivation has a positive and significant influence on debt and we can conclude that equality of access (social justice) is partly, at least, determined by ability to pay.

Figure 9 provides the relationship between the size of the debt ("debt intensity") and deprivation again for both metered and unmetered customers.[15] For metered properties, debt intensity is £150 per property in the least income-deprived areas, rising to £450 per property in the most deprived. For unmetered properties, average debt for the lowest five deciles is £325–£375 per property, increasing to over £500 for the most income-deprived areas. Although there is again a clear demonstration of the strong relationship between the size of the debt and the deprivation experienced by the people, the relationship differs between metered and unmetered households. In the most deprived areas, the

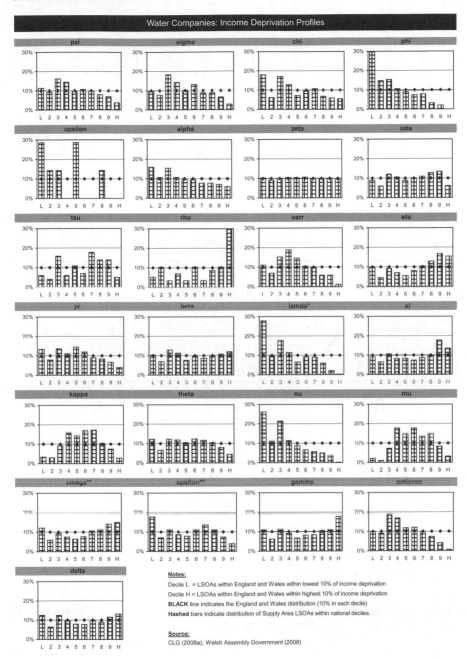

Figure 7 Income deprivation by deciles for all water company regions within the analysis.[15] (Original data from government statistics Welsh Assembly Government and DCLG).[19,20]

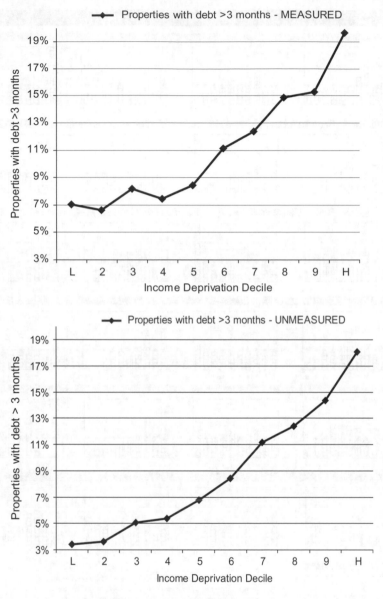

Figure 8   Proportions of households in debt at each deprivation level for metered and
unmetered households (national data).[15]

debt intensity is much the same whether households are metered or not, but in
the least deprived areas unmetered households typically have double the debt of
metered households. Figure 9, however, reveals that a smaller proportion of
unmetered households were in debt in the less deprived areas, and so the gross
debt (debt intensity × debt penetration) for an area may not significantly differ
between metered and unmetered households. Again the deprivation to debt

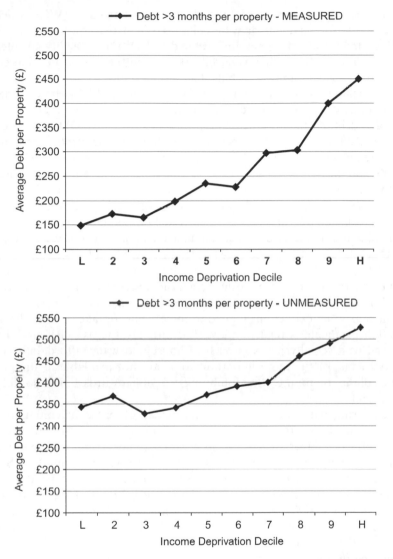

Figure 9 Average debts in debtor households at each deprivation level (national data).[15]

intensity relationship holds good across all companies examined and so appears to be a universal relationship within the England and Wales at least.

6.3 Developing a Socially just Response to Water Debt

There are many financial "tools" available to help payments for water: such as standing orders (SO), in which a static amount is paid periodically to the company; direct debit (DD), in which a variable amount is paid from a customer account as specified by the company; waters cards (WC), which are savings cards for payment of water bills to which the consumer can add a variable amount at any time; and water direct (WD), a facility to intercept benefits to settle outstanding water bills (limited in the proportion that can be intercepted and the numbers of bills that can be settled).

These facilities are not all readily available to deprived households. There is, for example, a strong inverse relationship between SO and DD arrangements and debt. As the proportion of households who have DD or SO facilities declines, debt increases. This arises as a direct consequence of affluence. The more affluent households hold bank accounts through which SO and DD operate and these act to safeguard payment (and, in effect, demonstrates that they have access to capital or credit). The water card is simply a savings device so that small, regular, flexible contributions towards a water bill can be made. In affluent (or, at least, less-deprived households) fewer than 10% of households use a water card, but this figure climbs in every water company such that in the most deprived areas water cards are held by between 15% and 35% of households. Water Direct is a facility to pay water bills directly from benefits payments. However, other creditors (electricity, gas, council) tend to be higher in the pecking order and, as the maximum top slice is 20% and the maximum deductions limited to three items, uptake of this scheme for water bill payments is relatively low. Encouraging the use of financial tools may well be a socially just approach to household water affordability in the UK, but some of the more important tools are not available to the poorest sections of the community, thus inhibiting progress towards a more socially just water regime.

Incentivising partners that have a particular expertise in communicating with customers is attractive, but is an approach that can hide methods that are clearly contrary to social justice. As customers begin getting into debt, water companies will send out progressively more severe reminders. Several companies have "subsidiaries" that appear to trade as independent debt-recovery bodies. Since these operate under implied threats of legal redress, and more, there is clearly a justice question arising out of this forceful approach. However, if the incentivised partner is the local authority that has a long-standing relationship with its citizens then the outcomes can be effective because the authority is a "trusted" debt collector. Alternatively, the companies can auction the debt to a third-party debt collector. The water company will receive only a small percentage of the debt outstanding, whilst the professional debt-collecting company is unrestrained in the methods it applies to debt recovery. If the collection methods employed use fear and threats without evaluating the ability of the "client" to settle the outstanding debt, then there is little or no social justice involved in such actions, or in a water company that has effectively acquiesced by passing debts to the debt-auction processes.

Demographic characterisation and customer tracking is not commonplace in (some might even say it is alien to) the water industry. In part this arises because the water industry does not have a contract with the customer – water is delivered as a right – and so the water companies have surprisingly limited information on their clients. In contrast, other utilities have a contract and so have much more detailed information. Delivering water as a right does, of course, confirm that water companies are operating in a very socially just manner in that they supply product to every household without evaluating ability to pay. If the companies could characterise customers, or even just track previous billing histories, they could readily identify which customers might have issues with water affordability and so could take early action to establish better contact, establish a more flexible tariff and billing structure, promote water cards and water direct membership and provide other support. Currently, customer problems are inevitably a surprise to the company.

The ultimate sanction is to cut off supplies. Whilst that is an option for energy utilities it is not one for water and, despite research indicating that disconnection does not appear to influence health,[17] nor should it be except in the most extreme circumstances! Cutting off of a fundamental life resource is not appropriate and is a measure that has no social justice whatsoever. But that requires us to consider how to manage those who simply won't pay, causing the financial burden to fall on all other customers – a clear injustice to the majority. The industry is now exploring, as evidenced by UKWIR research tender invitations, trickle supplies (or flow regulation devices) which will supply, inconveniently, a life-supporting supply of water, albeit at a very slow rate. There are a number of key questions that must be answered before considering the introduction of such a relatively draconian approach – an approach that until 6 months ago was rejected by the industry:

1. What are the health risks associated with such an approach?
2. What water volume per hour constitutes a trickle supply?
3. What undertakings are required for full flow resumption?
4. What approaches to debt management must have been exhausted before this option is exercised?

7 Social Justice and Water Futures

Thus far we have touched on social justice and the water customer and elements of corporate social justice in regard to water companies. This is only a sub-set of the stakeholders involved in water. If we consider the possible future climate and demography of the UK then there are several wider and perhaps more fundamental issues to address. The UK population is expected to exceed 65 million well before 2020. Tables 4 and 5 (derived from UK Climate Projections 2009; ref. 18) show the expected changes in summer temperature and summer rainfall, respectively, under a median emissions scenario by 2020. What does this mean in water resource terms and what are the consequences for social justice and water?

Table 4 Mean summer temperatures for UK regions in 2020 based on medium emissions scenarios.[18] The "wider range" is a term used to indicate the limits of possible, though less likely, climate shift.

Change in mean summer temperature (°C)
2020s Medium emissions scenario

Region	Probability level			Wider range	
	10%	*50%*	*90%*		
Channel Islands	0.5	1.4	2.5	0.4	2.5
East Midlands	0.5	1.4	2.5	0.4	2.5
East of England	0.5	1.4	2.5	0.5	2.6
Isle of Man	0.5	1.3	2.2	0.5	2.2
London	0.6	1.6	2.7	0.5	2.8
North East England	0.6	1.5	2.5	0.6	2.5
North West England	0.6	1.5	2.5	0.6	2.5
Northern Ireland	0.4	1.3	2.2	0.4	2.2
Scotland East	0.6	1.4	2.4	0.6	2.4
Scotland North	0.5	1.2	2.1	0.5	2.1
Scotland West	0.6	1.4	2.3	0.6	2.3
South East England	0.6	1.6	2.7	0.5	2.8
South West England	0.5	1.6	2.7	0.5	2.7
Wales	0.5	1.4	2.5	0.5	2.5
West Midlands	0.5	1.5	2.6	0.5	2.6
Yorkshire & Humber	0.5	1.3	2.3	0.4	2.3

Table 5 Mean summer precipitation for UK regions in 2020 based on medium emissions scenarios.[18] The "wider range" is the term used to indicate the limits of possible, though less likely, climate shift.

Change in mean summer precipitation (%)
2020s Medium emissions scenario

Region	Probability level			Wider range	
	10%	*50%*	*90%*		
Channel Islands	−31	−9	+17	−31	+22
East Midlands	−22	−6	+12	−22	+15
East of England	−24	−7	+12	−24	+15
Isle of Man	−21	−7	+9	−21	+10
London	−26	−7	+14	−26	+18
North East England	−19	−6	+8	−19	+9
North West England	−23	−8	+9	−23	+10
Northern Ireland	−17	−5	+7	−17	+10
Scotland East	−17	−6	+7	−17	+8
Scotland North	−15	−4	+7	−15	+8
Scotland West	−17	−6	+7	−17	+8
South East England	−26	−8	+14	−26	+18
South West England	−27	−8	+14	−27	+18
Wales	−23	−7	+11	−23	+14
West Midlands	−23	−7	+12	−23	+15
Yorkshire & Humber	−24	−8	+10	−24	+11

In water resource terms, 2020 is the immediate future. Most water companies would not be confident that they could get agreement, permissions and appropriate designs in place to deliver new supply infrastructure in that time frame, even starting now (2010). A reduction of 5 to 10% in summer rainfall across the UK – in association with a 1 to 2% rise in mean summer temperatures – would stimulate domestic water demand slightly, but would drive up agricultural demand significantly at a time of lowered river levels.

Let us assume that the water for people will still be delivered, *i.e.* that industrial and domestic demand will be satisfied. The question that arises is whether there would be sufficient water in all streams at all times to permit irrigation abstractions and to safeguard the aquatic ecosystem. We cannot assume that we will continue to obtain food (and thus associated water) from overseas, partly because food security is a growing issue and partly because concerns have been raised (see Chapter 4 in this volume) about the water hidden in food (variously known as "latent water", "water footprint", *etc*). We may well then be faced with a situation in which not all water demands can be met and selection and prioritisation will be required. To abandon good ecological status in river reaches would contravene the Water Framework Directive and, more important from the perspective of this chapter, would compromise environmental status and perhaps justice if these impacts affect the wellbeing of people. Would we compromise agricultural productivity and thus food security by limiting agricultural abstractions? This would impact primarily on arable crop areas which may have to switch to grazing systems – a highly inefficient food source. Both the farming and the food users (*i.e.* all of us) would have to bear the productivity losses and price rises, price rises which would, of course, in the longer run impact more heavily on food availability in the third world. The scale, both spatially and in severity, of the social justice implications is very concerning. To avoid being too academic and sitting on the fence our view is: (i) that good ecological status cannot have a sacrosanct protected status, and (ii) that significant changes to winter storage, despite the impact of reservoirs on sensitive landscapes, will have to be implemented.

Clearly such extreme responses would have to be paralleled by major water efficiency actions at both the domestic scale and by improved national water resources management, but without such bold steps we will face water shortages that impact on all people and sectors. Of course, that would be equity but not the equity envisaged when social justice and water is raised.

References

1. P. Cullen, B. Hoose and G. Mannion, *Catholic Social Justice: Theological and Practical Explorations*, T. & T. Clark, Continuum, London, UK and New York, NY, USA, 2007.

2. Social Justice, Dictionary.com's 21st Century Lexicon, Dictionary.com, LLC; http://dictionary.reference.com/browse/social justice, accessed 30/03/2010.
3. Department of Transportation, *Environmental Justice Program Goals*; http://www.dotcr.ost.dot.gov/asp/EJ.asp?print = true#TOC, accessed 28/03/2010.
4. M. Black, *Learning What Works: a 20 Year Retrospective Review on International Water and Sanitation Cooperation*; http://www.un.org/esa/sustdev/sdissues/water/InternationalWaterDecade1981-1990_review.pdf, accessed 22/03/2010.
5. WHO/UNICEF, *Joint Monitoring Programme for Water Supply and Sanitation,* 2010; http://www.wssinfo.org/essentials/water2.html, accessed 30/03/2010.
6. E. Esrey, Water, waste, and well-being a multicountry study, *Am. J. Epidemiol.*, 1996, **143**(6), 608–623.
7. S. Sasaki, H. Suzuki, Y. Fujino, Y. Kimura and M. Cheelo, Impact of drainage networks on cholera outbreaks in Lusaka, *Zambia Am. J. Public Health*, 2009, **99**(11), 1982–1987.
8. L. Fewtrell, R. B. Kaufmann, D. Kay, W. Enanoria, L. Haller and J. M. Colford Jr, Water, sanitation and hygiene interventions to reduce diarrhoea in developing countries: a systematic review and meta-analysis, *Lancet Infect. Dis.*, 2006, **5**, 42–52.
9. World Health Organisation, *Dracunculiasis (Guinea Worm) Eradication*, 2010; http://www.afro.who.int/en/divisions-a-programmes/ddc/communicable-disease-prevention-and-control/programme-components/dracunculiasis-guinea-worm-eradication.html, accessed 30/03/2010.
10. K. Bates and R. Swan, *Through the Eye of Katrina: Social Justice in the United States,* Carolina Academic Press, Durham, NC, USA, 2007.
11. Environment Agency, *The Costs of the Summer 2007 Floods in England*, Project *SC070039*, Environment Agency, Bristol, UK, 2010.
12. K. Bakker, *Privatising Water in England and Wales: Water Pricing and Equity*, AWRA/IWLRI-University of Dundee International Specialty Conference, 2001; http://www.awra.org/proceedings/dundee01/Documents/Bakker.pdf, accessed 30/03/2010.
13. DOE, *Review of the Water Industry in England and Wales,* HMSO, London, 1976.
14. S. E. Gaines, The polluter pays principle: from economic equity to environmental ethos, *Texas Int. Law J.*, 1991, **26**, 423–4624.
15. M. Clarke, P. Boden and A. McDonald, *Debt Collection Performance and Income Deprivation*, UK Water Industry Research Report 09/CU/04/6, UKWIR, London, UK, 2009.
16. House of Lords, *Water Management, Science and Technology Committee, HL Paper 191 – I and II*, 2006.
17. L. Fewtrell, D. Kay, J. Dunlop, G. O'Neill and M. Wyer, Infectious diseases and water-supply disconnections, *The Lancet*, 1994, **343**, 1370.

18. *UK Climate Impacts Programme*, 2010; http://www.ukcip.org.uk/index.php?option = com_content&task = view&id = 357, accessed 30/03/2010.

19. Communities and Local Government, *Indices of Deprivation 2007*, 2008; http://www.communities.gov.uk/communities/neighbourhoodrenewal/deprivation/deprivation07/, accessed 14/04/2010.

20. Welsh Assembly Government, *Welsh Index of Multiple Deprivation (WIMD) 2008. Summary Report*, 2008; http://new.wales.gov.uk/statsdocs/compendia/wimd08/summary/wimd08sumintroe.pdf?lang = en, accessed 14/04/2010.

Safe Management of Chemical Contaminants for Planned Potable Water Recycling

STUART KHAN

ABSTRACT

Planned potable water recycling has become an important strategy for managing drinking water supplies in a growing number of countries including Australia, the USA, Namibia, Singapore and some parts of Europe. This involves the advanced treatment of municipal wastewaters (treated sewage) prior to reuse by supplementation of drinking water supplies. In order to ensure the protection of public health, careful assessment and management of trace chemical contaminants is essential. Important chemical contaminants include pharmaceuticals, personal care products, steroidal hormones, pesticides, industrial chemicals, perfluorochemicals, antiseptics and disinfection by-products. Monitoring of key toxic chemicals of concern is important and can be supplemented by techniques for direct toxicological assessment of whole water samples. Furthermore, monitoring of a limited range of identified indicator chemicals and surrogate parameters can be an effective means for confirming ongoing treatment performance of individual unit treatment processes. This chapter provides a description of some of the most significant planned potable water recycling schemes currently in operation or under development around the world. It highlights the various approaches used in the assessment of chemical safety for these schemes, and hence the alternatives that are available for the assessment of future potable water recycling schemes.

Issues in Environmental Science and Technology, 31
Sustainable Water
Edited by R.E. Hester and R.M. Harrison
© Royal Society of Chemistry 2011
Published by the Royal Society of Chemistry, www.rsc.org

1 Introduction: Planned Potable Water Recycling

In an increasing number of major urban centres around the world, combinations of rapid population growth and extended periods of drought have led to a growing imbalance between potable water supply and demand. The long-term sustainability of water supplies in these regions will require an effective combination of demand management and resource enhancement or supplementation. One approach that is growing in application is planned potable water recycling.

Planned potable recycling of municipal wastewater refers to the purposeful augmentation of a potable water supply (surface water or groundwater) with highly treated reclaimed water derived from conventionally treated municipal effluents. In "indirect" potable recycling (IPR) schemes, the reclaimed water is returned to an environmental "buffer" such as a river, lake or aquifer, where it mixes with other environmental waters before being re-extracted for drinking water treatment and potable use.

So-called "unplanned IPR" or "incidental IPR" is common throughout many parts of the world. This involves situations where municipal drinking water supplies are taken from waterways that are impacted by wastewater discharge, usually by upstream towns or cities. The distinction between "unplanned IPR" and "planned IPR" may not always be clear, but planned IPR usually involves an acknowledgement that reclaimed effluent is an important component of the potable water supply and thus the selected water treatment processes, risk assessment and risk management activities explicitly reflect this fact. In many cases, planned IPR involves recycling of the reclaimed water back into the same potable water supply, rather than that of a downstream community.

IPR differs from direct potable recycling where highly treated reclaimed water is introduced "directly" into the drinking-water distribution system without delay or further treatment. Windhoek in Namibia is currently the only community that has a direct potable recycling scheme. Major population centres with established planned IPR schemes include Singapore, Orange County in California (USA), the Upper Occoquan region in Virginia (USA) and the Veurne Region in north western Belgium.

In Singapore, between 1 and 2% of treated municipal wastewater is returned to the island's raw water reservoirs at Kranji, Bedok and Seletar. Orange County in California has operated an aquifer-injection IPR scheme since 1976. This scheme, now known as the Orange County Groundwater Replenishment System, was expanded in 2007 to achieve a supply capacity of over 260 megalitres of recycled water to the drinking water aquifer per day. The Upper Occoquan Sewage Authority in Virginia, USA produces close to 200 megalitres of recyclable water per day, which is discharged into the Upper Occoquan reservoir to supply nearly one million people in the area around Washington DC. The Torreele/Wulpen IPR scheme in the Belgian Veurne Region recharges a shallow groundwater aquifer with a production capacity of 2500 megalitres per year. This is equivalent to around 40% of the local potable water demand.

The construction of an IPR scheme known as the Western Corridor Recycled Water Project (WCRWP) was completed in Queensland, Australia in 2009. The WCRWP was designed to use the vast majority of treated municipal effluent produced in South East Queensland and previously discharged to the marine environment. This municipal effluent is now collected from six waste-water treatment plants, and then subjected to advanced water treatment at three new advanced water treatment plants around Brisbane. Some of the advanced-treated water is now used to supply cooling water for two coal-fired power stations. The remaining water will be used to recharge Brisbane's main drinking water reservoir, Lake Wivenhoe. However, this IPR aspect is expected to be initiated only once Brisbane's combined storage levels drop below 40% of capacity. The reclaimed water will mix with other environmental sources of water in Lake Wivenhoe, before undergoing further treatment at Brisbane's main drinking water treatment plant and then distribution as potable water.

All of these IPR schemes incorporate conventional municipal sewage treat-ment processes followed by a variety of advanced water treatment processes. The approach pioneered at Orange County, California incorporates an inte-grated membrane treatment system (microfiltration or ultrafiltration followed by reverse osmosis). Further treatment at the Orange County scheme is by advanced oxidation using ultraviolet irradiation and hydrogen peroxide. This approach has been adopted for a number of subsequent schemes including the WCRWP in Queensland, Australia. Other schemes, such as the Upper Occoquan scheme rely on more conventional treatment processes including adsorption of chemicals to activated carbon and disinfection using chlorine and/or ozone.

The design philosophy of all of these schemes is in accordance with the "multiple barriers" approach, incorporating numerous largely independent treatment steps operating under a risk management framework. The need for such a focus on risk management derives from the inherently poor qualities of the source waters used for IPR applications. The source waters – municipal wastewater effluents – contain a wide range of pathogenic microorganisms and toxic chemicals. The careful and effective management of human health risks associated with these substances is paramount for any IPR scheme. The following sections of this chapter describe some of the key approaches that have been employed and promoted for the safe management of chemical contaminants in planned IPR schemes.

2   Chemical Contaminants in Potable Water Recycling

Chemical hazards in municipal wastewaters consist of a wide range of naturally occurring and synthetic organic and inorganic species. They include industrial chemicals, chemicals used in households, chemicals excreted by people and chemicals formed during wastewater and drinking-water treatment processes.

Depending on the catchment area, and the extent of the trade waste pro-gramme to control chemicals at the wastewater source, a very wide range of synthetic industrial chemicals are often measurable in urban municipal wastewaters. Examples include plasticisers and heat stabilisers, biocides, epoxy resins, bleaching chemicals and by-products, solvents, degreasers, dyes, chelating agents, polymers, polyaromatic hydrocarbons, polychlorinated biphenyls and phthalates.

Volatile organic compounds (VOCs) are widely used as industrial solvents. Many are constituents of petrochemical products, and a number of halogenated compounds may be formed as by-products of chlorine disinfection.

Heavy metals may be present in municipal wastewaters as a result of industrial discharges to sewers. The presence of some metals can be related to geological and soil conditions in the potable water catchment.

Pesticides may enter municipal wastewater systems by a variety of means including stormwater influx and illegal direct disposal to sewage systems. Some leftover household pesticides are known to be disposed of *via* municipal sewers. Additional routes, of unknown significance, include washing fruit and vegetables prior to household consumption; insect repellents washed from human skin; flea-rinse shampoos for pets; and washing clothes and equipment used for applying pesticides.

Algal toxins such as microcystins, nodularins, cylindrospermopsin and saxitoxins are all produced by freshwater cyanobacteria (blue-green algae). Under suitable conditions, cyanobacteria may grow in untreated or partially-treated wastewaters, producing these and other toxins. Numerous algal toxins have been implicated as having serious impacts on human and animal health by the consumption of contaminated water.

Disinfection by-products are formed by reactions between disinfection agents and other constituents of water. High initial concentrations of organic components may lead to excessive production of disinfection by-products. The vast majority of the compounds of concern originate from chlorine-based disinfectants. However, some (such as formaldehyde) can be produced by other oxidising disinfectants such as ozone. Some more recent by-products of concern include bromate and epoxides (from ozone treatment) and nitrosodimethylamine (NDMA), predominantly from chloramination.

Radionuclides may enter sewage by natural run-off, or as a result of medical or industrial usage. In most parts of the world, radium is a constituent of bedrock and hence a natural contaminant in groundwater. In some cases, radium is removed from drinking waters by coagulation and the concentrated sludge is transferred to sewage systems.

Pharmaceuticals (and their active metabolites) are excreted to sewage by people, as well as direct disposal of unused drugs by households. Since pharmaceuticals are designed to instigate biological responses, their inherent biological activity and the diverse range of compounds identified in sewages (and the environment) have been cause for considerable concern during the last decade.

Natural and synthetic steroidal hormones such as oestradiol, oestrone, ethinylestradiol and testosterone are also excreted to sewage by people. During the last two decades, steroidal hormones have been widely implicated in a range of endocrinological abnormalities in aquatic species impacted by sewage effluents.

   Antiseptics such as triclosan and triclocarban are commonly used in face
washes and anti-gum-disease toothpaste. They are increasingly being used in a
wider range of household products including deodorants, antiperspirants,
detergents, dishwashing liquids, cosmetics and anti-microbial creams, lotions,
and hand soaps.

   Perfluorochemicals, such as perfluorooctanoic acid (PFOA) and perfluoro-
octane sulfonate (PFOS), are persistent and toxic chemicals that have recently
emerged as drinking water contaminants of concern. These perfluorochemicals
are highly water soluble and are used in the production of water- and stain-
resistant products including cookware and clothing, as well as in firefighting
foams. They also arise from the breakdown of fluorotelomer alcohols, which
are widely used in consumer products such as greaseproof food wrappers and
stain-resistant carpet treatments.

   A further important group of emerging environmental contaminants of
concern is nanoparticles or nanomaterials. The toxicological concerns for
nanoparticles are related not only to their chemical composition, but also to
physical parameters including particle size, shape, surface area, surface chemistry, porosity, aggregation tendency and homogeneity of dispersions. As such, it
is increasingly recognised that traditional techniques for toxicological and
ecotoxicological evaluation of chemical substances are not well applied to the
evaluation of nanoparticles.

3   Chemical Risk Assessment and Potable Water Recycling

Risk assessment provides a systematic approach for characterising the nature
and magnitude of risks associated with environmental health hazards. Formalised requirements for risk assessment for drinking water were introduced in
the United States with amendments to the US Safe Drinking Water Act in
1974. These amendments required improved estimates of exposure to potential
hazards for risk management purposes. In 1983 the US National Research
Council published what became known as the "red book",[1] which provided a
formalised set of steps to be taken for assessing risks to human health by
chemicals from environmental and other sources. These were:

1. Problem Formulation and Hazard Identification – to describe the
   human health effects derived from any particular hazard (*e.g.* acute
   toxicity, carcinogenicity, *etc.*).
2. Exposure Assessment – to determine the size and characteristics of the
   population exposed and the route, amount and duration of exposure.
3. Dose–Response Assessment – to characterise the relationship between
   the dose exposure and the incidence of identified health impacts.
4. Risk Characterisation – to integrate the information from exposure,
   dose response, and health interventions in order to estimate the magnitude of the public health impact and to evaluate variability and
   uncertainty.

These steps have evolved into a general framework now used by environmental health agencies throughout the world to assess risks posed by environmental human health hazards including chemicals and microbial organisms.

Risks associated with chemical contaminants in water are highly variable depending on the precise chemical species. Some chemicals may be acutely toxic, meaning that they impart toxic effects in a short period of time subsequent to a single significant dose. However, most toxic chemicals present chronic health risks, meaning that long periods of exposure to small doses can have a cumulative effect on human health.

The toxicological evaluation of chemical contaminants is normally undertaken by animal testing, most commonly rodents, including mice, rats and hamsters. The effects of chronic (lifetime) or sub-chronic (up to 10% of the lifespan) exposure, rather than acute effects from brief exposure to high doses, are considered to be most applicable for drinking water supplies. In rodents, this corresponds to about two years for chronic exposure and two to thirteen weeks for sub-chronic exposure experiments. For assessment of developmental effects, studies involving exposure during gestation are relevant. Chemicals may be administered to animals by drinking water, diet or by "gavage", which involves the delivery of a single large dose dissolved in oil or water through a tube into the stomach. In some cases, it may be appropriate for exposure to be assessed by inhalation, dermal adsorption, or intraperitoneal, intravenous, or intramuscular injection.

Animal toxicological studies for non-carcinogenic effects typically involve examination of mortality rate, body weight, organ weights, microscopic appearance and enzyme activities. Among the organs most frequently monitored are liver and kidney because of their roles in metabolism and excretion of toxic chemicals. Increasing attention is being given to studies of teratogenicity and developmental and reproductive toxicology as these periods of development are now known to be the most sensitive life stages for many chemicals. For the investigation of carcinogenic effects, a wide variety of organs and tissues are examined for tumours. The incidence of malignant and benign tumours in the experimental group is compared to a background incidence in a control group.

Quantification of the level of toxicity imparted by specific chemicals is customarily undertaken by the development of a "dose–response" curve. The dose–response curve defines the relationship between the level of exposure (the dose) of a chemical and incidence of (usually negative) health impacts.

For most toxic effects of chemicals, a clear dose–response relationship exists, with the proportion of responses increasing over a certain dose range. For many toxic effects there is a "threshold" dose, below which no toxic effects are observed. However, for some effects, particularly cancer which occurs through a genotoxic mode of action, it is assumed that exposure to any dose results in some level of risk; thus there is no threshold below which no risk exists.

Toxicological investigations for non-carcinogenic ("threshold") studies are usually designed to enable the identification of either the highest dose at which no adverse effects are observed (the No Observed Adverse Effects Level, NOAEL) or else the lowest dose at which adverse effects are observed (the Lowest Observed Adverse Effect Level, LOAEL). NOAELs and LOAELs are conventionally determined in units of milligrams of the substance per kilogram of body mass per day ($mg\,kg^{-1}\,day^{-1}$).

In the determination of safe exposure levels, uncertainty factors (sometimes called safety factors) are usually applied. These are used to account for uncertainty derived from incomplete toxicological databases, the use of sub-chronic studies to derive chronic effects, the use of a LOAEL in place of a NOAEL, the extrapolation of animal studies to human impacts, and to account for variability amongst the human population. Each of these uncertainty factors is conventionally applied as a value of either 10 (a full log_{10} unit) or 3 (approximately half a log_{10} unit), up to a maximum product of 10 000 on the basis that exceedance of this value would imply a degree of uncertainty which would no longer support a robust scientific decision.[2] For some cases, the US EPA has recommended limiting uncertainty factors to a maximum of 3000 to avoid the overly conservative result of applying the maximum value (10) to four or more sources of uncertainty.[3]

Safe drinking water concentrations for non-carcinogenic chemicals can be calculated from appropriate NOAEL or LOAEL data and relevant uncertainty factors by applying a number of assumptions regarding exposure to drinking water. Drinking water guideline values or standards are commonly calculated according to Equation (1) or variations of it.[4–6]

$$\text{Safe Drinking Water Concentration (mg/L)} = \frac{\text{POD (NOAEL or LOAEL)} \times \text{BW} \times \text{PF}}{\text{IR} \times \text{UF}} \tag{1}$$

where:

BW = average body weight of an adult (commonly 70 kg).

PF = a proportionality factor to account for the assumed contribution of drinking water to the total exposure to the chemical (typically 1 or 0.1).

IR = the estimated maximum drinking water ingestion rate by an adult ($2\,l\,day^{-1}$).

UF = one or more uncertainty factors (usually values of 10 or 3 to a maximum product of 3000; ref. 2).

Toxicological assessments for carcinogenic (and some other) chemicals generally assume that there is no "threshold level" at which there is no increased risk of detrimental impact.[7] Accordingly, it is assumed that there is no "safe level" of exposure and the calculation of NOAELs or LOAELs is not applicable.

Dose–response curves for carcinogenic chemicals are customarily characterised in terms of a "Cancer Slope Factor" (CSF), in which cancer risk per lifetime daily dose is given in inverse exposure units of $(mg\,kg^{-1}\,day^{-1})^{-1}$. The carcinogenic risk then is assumed to be linearly proportional to the level of exposure to the chemical, with the CSF defining the gradient of the dose–response relationship as a straight line projecting from zero exposure–zero risk. A sharper gradient, defined by a higher CSF, indicates a more potently carcinogenic chemical, leading to increased cancer risk for any identified level of exposure.

The CSF for a specific carcinogen may be determined from human epidemiological studies or (more commonly) from chronic animal carcinogenicity assays. It is then used to calculate the probability of increased cancer incidence over a person's lifetime – the so-called "excess lifetime cancer risk"– as a function of exposure to the chemical.

Since it is assumed that there is no "threshold dose" for most carcinogenic chemicals, the assumption that any level of exposure incorporates some level of risk means that the "safe" level of exposure must either be defined as zero exposure or else in terms of an identified "tolerable level of risk". On this basis, the US EPA has set non-enforceable Maximum Contaminant Level Goals (MCLGs) for carcinogenic chemicals in drinking water as zero. The enforceable standard, the Maximum Contaminant Level (MCL), is determined as the level that may be achieved with the use of the best available technology, treatment techniques, and other means which EPA finds are available, taking cost into consideration.

Some international agencies have identified a tolerable level of excess lifetime cancer risk, usually 10^{-4}, 10^{-5} or 10^{-6}. Having defined the tolerable level of risk, a safe drinking water concentration may then be calculated according to Equation (2) or variations of it.[4 6]

Equation (2) Calculation of safe drinking water concentration for carcinogenic chemicals.

$$\text{Safe Drinking Water Concentration (mg/L)} = \frac{\text{Risk level} \times \text{BW} \times \text{PF}}{\text{CSF} \times \text{IR}} \tag{2}$$

where:

Risk Level = tolerable risk level (usually 10^{-4}, 10^{-5} or 10^{-6}).

BW = average body weight of an adult (commonly 70 kg).

PF = a proportionality factor to account for the assumed contribution of drinking water to the total exposure to the chemical (typically 1 or 0.1).

IR = the estimated maximum drinking water ingestion rate by an adult ($2 \, \text{l day}^{-1}$).

Established drinking water guidelines and standards are generally limited to "traditional" drinking water contaminants, known to occur in contaminated surface water from conventional supplies.[4,5] These include a range of pesticides and industrial chemicals, but not trace chemicals associated with discharges from municipal wastewater effluents. Accordingly, contaminants of concern, such as pharmaceutical residues, personal care products, household chemicals and steroidal hormones, are not commonly subject to specific regulation in drinking water.

Nonetheless, safe drinking water concentrations of many of these chemicals may potentially be derived from the same toxicological considerations used for the establishment of current guidelines and standards. For example, suitable toxicological data for determining safe drinking water concentrations have been reported for 26 active pharmaceutical ingredients based on various end-points.[8] The authors of this study stated that, for pharmaceutical substances, the therapeutic effect usually occurs at a dose considerably below that expected to result in toxicity and thus a large proportion of the NOAEL or LOAEL toxicological data used were based on therapeutic effects or minor side-effects such as sensitivity to human intestinal microflora. However, others have reported that for many pharmaceuticals, the most toxic endpoint is not the therapeutic endpoint, but rather it is a side-effect, such as carcinogenicity.[9] In the absence of other relevant toxicological data, the Australian Guidelines for Water Recycling have adopted the use of lowest therapeutic dose in place of a traditional LOAEL.[10]

In some cases, however, there is insufficient toxicological data to set a drinking water guideline value by the conventional method. In such situations, the "threshold of toxicological concern" (TTC) concept can help set an interim guideline while further toxicological information is gathered.[11,12] The TTC model largely relies on carefully compiled toxicological databases of acute, sub-chronic and chronic NOAELs from rodent studies to which a safety factor of 100 has been applied for extrapolation to humans. The concept was originally developed to prioritise toxicity testing of food additives and food contact materials, but the methodology can be applied to other occupational and environmental settings, including drinking water contaminants.[13,14] This approach has been adopted by the Australian Guidelines for Water Recycling as a means of identifying con-servative upper-bounds for safe concentrations of some chemicals as an interim measure until sufficient toxicological data become available.[10]

4   Relative Risk

The techniques described above are aimed at the quantification of human health risks from chemicals in absolute terms according to known dose–response relationships. However, in many cases, absolute quantification of risks is not possible or not practical due to a lack of relevant toxicological data. In such cases a relative risk assessment approach can be useful.

Relative risk assessment can compare the degree of exposure to a chemical from a new recycled water source, relative to the exposure from a pre-existing water source. Such an approach can be useful for demonstrating reduced or increased risks associated with exposure to a particular contaminant in the absence of quantitative toxicology/infectivity data.

Relative risk assessment was the key approach adopted for a direct potable reuse study undertaken in San Diego during the 1990s.[15] The primary objective of that study was to assess whether a pilot-scale advanced water treatment plant could reliably reduce contaminants of concern to levels such that the health risks posed by an assumed potable use would be no greater than those associated with the present water supply. Over a period of three years, water quality testing was undertaken for a series of pathogenic organisms, potentially toxic chemicals, as well as screening for mutagenicity and bioaccumulation of the chemical mixtures present in the two water supplies. The results of this research indicated that the advanced water treatment plant could reliably produce water of equal or better quality than that of the present water supply.[15]

5 Direct Toxicity Testing

Even if the operators of a water recycling system could identify all of the organic chemical components in the specific municipal effluent, there would be scant toxicological data available for most of them and thus little basis for assigning risks. Other important limitations of chemical species monitoring are that the full additive toxicity of a large number of chemicals, each present at very low concentrations may not be identified unless each of the individual species is analysed for and determined to be present at concentrations greater than analytical detection limits. Finally, there is widespread current concern that the toxicity of complex mixtures is poorly understood and in some cases may amount to more (or less) than simply additive impacts from the summation of each of the contributing chemical species.

To overcome the limitations of chemical analysis, many scientists have suggested that direct toxicological testing assessment of recycled water may be the most effective way to ensure the water's chemical safety.[16] Toxicological testing involves collecting whole water samples and subjecting these to tests for a range of toxicological assays. Toxicological endpoints may include testing for mutagenic activity, carcinogenic activity, hormonal activity such as estrogenicity, or various forms of acute toxicity.

5.1 *In vivo Toxicity Testing*

The most comprehensive approach to toxicity testing is considered to be live animal testing with organisms such as rats, mice and fish. One particular outbred line of mice has been used extensively for carcinogenesis experiments. These strains are known as SENCAR mice, derived from SENsitivity to CARcinogenisis. SENCAR mice are used to test for the presence of carcinogens and/or promoter agents to induce carcinogenesis. Other live animal testing may include monitoring for sub-chronic toxicity (leading to death or other endpoint indicators) or fetotoxicity. Fish biomonitoring may also be employed to test for effects of bioaccumulating chemicals as well as a range of toxic endpoints.

A comprehensive two-year health effects study was undertaken in Denver, Colorado to investigate chronic toxicity and oncogenicity effects of the reclaimed water using rats.[17] These tests were conducted using 150-fold and 500-fold reclaimed water concentrates. Denver's current drinking water was used as a negative control since it is derived from a relatively protected source. Seventy male and 70 female Fischer 344 rats were supplied with concentrates of one of three kinds of water. These were drinking water, reverse osmosis-treated reclaimed water or ultrafiltration-treated reclaimed water. The parameters evaluated in this study included clinical observations, survival rate, growth, food and water consumption, haematology, clinical chemistry, urinalysis, organ weights, gross autopsy and histopathological examination of all lesions, major tissues and organs.

Fish biomonitoring experiments were undertaken to assess a pilot potable recycling plant in San Diego, California (the San Diego AQUA III plant). The aim was to provide information on chronic exposure to trace contaminants that accumulate in tissue but are not known to be identified by genetic toxicity screening bioassays.[18] Juvenile fathead minnows (*Pimephales promelas*) were exposed to AQUA III plant effluent, drinking source water or laboratory-control water in flow-through aquaria. The biomonitoring measurements were survival and growth, swimming performance, and trace amounts of 68 base/neutral/acid extractable organics, 27 pesticides and 27 inorganic chemicals found in fish tissues after exposure.

Fish studies were also undertaken in Singapore as a component of the NEWater Study.[19] The purpose was to assess long-term chronic toxicity as well as the estrogenic potential (reproductive and developmental). The orange-red strain of the Japanese medaka fish (*Oryzias latipes*) was selected for the study due to the availability of an extensive biological database for this species.

A more recent study undertaken in Orange County California has demonstrated the use of live fish for a long-term, on-line flow-through bioassay used to evaluate recycled water quality.[20] The approach that was used in this study included monitoring for the effects of endocrine-disrupting, tissue-altering and carcinogenic compounds.

5.2   In vitro Toxicity Testing

In vitro toxicity tests are tests performed at the molecular or cellular level in the laboratory. Examples of molecular endpoints include binding to specific biological receptors or induction of particular biomolecular pathways, while cellular events could include cell death, maturation or growth. *In vitro* assays can be based on human cells, thus eliminating the inter-species predicament of *in vivo* testing. *In vitro* tests can also detect biological effects at much lower, environmentally relevant concentrations, often below detection limits of chemical analysis and *in vivo* testing.[21] There are, of course, also limitations to *in vitro* bioassays, in particular the lack of metabolism and transport mechanisms that may modulate toxicity in whole organisms.

In vitro assays were developed for screening purposes and there is still much debate about their ability to predict whole organism effect.[16] Nevertheless, *in vitro* bioassays are well-suited to monitoring of water quality, as they are significantly faster and cheaper than *in vivo* exposures, are amenable to high-throughput screening, and allow the generation of relatively rapid toxicology data without the need for ethically and financially expensive whole animal experimentation.[22]

In recent years, there has been a move towards standardising the various *in vitro* techniques available, with the creation of the European Centre for the Validation of Alternative Methods (ECVAM) in 1991 and the US National Toxicology Program Interagency Centre for the Evaluation of Alternative Toxicological Methods (NICEATM) in 1998. These two programs have generated thoroughly validated alternative methods using *in vitro* toxicity tests for some toxic endpoints.

Chemicals can be considered depending on their potential for subsequent effects in exposed humans and wildlife, and *in vitro* toxicity tests exist for a variety of endpoints including cytotoxicity, immunotoxicity, genotoxicity and mutagenicity, neurotoxicity, teratogenicity and fetogenic changes, as well as various forms of endocrine disruption.[23]

Each type of bioassay has its advantages and limitations, and no single assay can provide a complete assessment of the biological activity of a sample. Therefore a battery of bioassays is required to rigorously assess the biological effects potential of a sample.

A five-year toxicological study was initiated in 1978 at the Montebello Forebay Groundwater Replenishment Project, which is an IPR scheme located within the Central Groundwater Basin in Los Angeles County.[24] At the time of the study, recycled water comprised around 16% of the total inflow to the groundwater basin. The toxicological study sought to detect, isolate, characterise and, if possible, trace the origins of any previously unidentified carcinogens in the recycled water sources (three STP effluents) and well waters. The Ames test and *Salmonella* tester strains (TA98 and TA100) were used to screen for mutagenic organics in 10 000 to 20 000-fold concentrates of reclaimed water prior to groundwater replenishment, stormwater, imported water, and also in chlorinated and unchlorinated groundwaters.[24]

In the 1980s, the Potomac Estuary Experimental Water Treatment Plant (EEWTP) was constructed adjacent to a conventional sewage treatment plant on the Potomac Estuary in Washington DC. The US Army Corps of Engineers undertook a research program to determine the feasibility of using the EEWTP to produce potable water as a potential source for the city.[25] Based on potential future water use plans, an influent mix of 50% estuary water and 50% nitrified secondary effluent was selected for further treatment at the EEWTP. The blended water underwent treatment by a series of processes that are generally considered to be conventional drinking water treatment processes (such as flocculation, sedimentation and disinfection), as well as some additional processes including filtration and granular activated carbon adsorption. Three overall treatment trains were tested and these were compared to current water supplies from three local conventional drinking water treatment plants. Two short-term *in vitro* toxicology tests were selected to characterise the produced and conventional waters. These tests were undertaken with concentrated

(150-fold) organic extracts used in the Ames *Salmonella*/microsome test, and a mammalian cell transformation test.

A health effects study was undertaken to compare the San Diego AQUA III plant effluent with the city's present raw water supply.[15,26,27] The study included screening for mutagenicity and bioaccumulation of the chemical mixtures present in the conventional water produced by the proposed IPR scheme. Genetic toxicity and carcinogenicity testing were undertaken in a short-term study using four separate bioassay systems. These were the Ames Assay, Micronucleus Test, 6-Thioguanine Resistance Assay and Cellular Transformation Assay.

A pilot advanced water treatment plant, known as the Tampa Water Resource Recovery Project (TWRRP) was operated in Tampa, Florida from 1987 to 1989.[28] The TWRRP was originally operated using chlorine as the final disinfectant, but this was replaced with ozone since the results of Ames testing indicated that ozone-disinfected product waters were less mutagenic. For similar reasons, the treatment train using GAC was selected for toxicological testing based on preliminary screening using the Ames assay. The performance of TWRRP was assessed in comparison Tampa's current water supply. Concentrated extracts of these water sources were used to prepare doses for toxicological testing at up to 1000 times the potential human exposure of a 70 kg person consuming 2 litres of water per day. Eight different toxicological tests were conducted to assess potential genotoxicity (Ames and sister chromatid assays), carcinogenicity (strain A lung adenoma and SENCAR mice initiation-promotion studies), fetotoxicity (teratology in rats and reproductive effects in mice), and sub-chronic toxicity (90-day gavage studies in mice and rats).[29]

6   Indicator Chemicals and Surrogate Parameters

An important task in designing a monitoring program for a water treatment system is to determine which parameters should be monitored. One approach is to identify key contaminants of concern from a toxicological point of view. For some specific chemicals, there may be a valid justification for doing this. However, this approach can never be sufficiently comprehensive to ensure sufficiently low concentrations of all potential toxic contaminants. Toxicity testing may help to address this limitation, but toxicity also suffers from some significant practical limitations, including the resources and time required to undertake testing, as well as an inability to test the very large number of potentially toxic endpoints, which may be targeted by various chemicals.

An additional approach for a chemical monitoring program is to aim to confirm that the water treatment processes are operating correctly and performing sufficiently well to produce water of an acceptable quality. This approach implies that there is some known relationship between the observed performance of a water treatment process and the produced water quality. Accordingly, there is a need to validate such assumptions where they are made.

One approach that has recently been validated in concept is the use of indicator compounds and surrogate measures to assess and monitor water treatment process performance.[30] In this context, an *indicator compound* is an individual chemical, occurring at a quantifiable level, which represents certain physicochemical and/or biodegradable characteristics of a larger group of trace constituents. The representative characteristics must be relevant to the transport and fate of the indicator compound, as well as the larger represented group of constituents, providing a conservative assessment of removal. A *surrogate parameter* is a quantifiable change of a bulk parameter that can serve as a performance measure of individual unit processes or operations regarding their removal of trace compounds.

The ultimate advantage of the "indicators and surrogates" approach is that it requires only a limited set of analytes to be measured in order to assess and monitor treatment processes for an assumed much larger group of chemicals. Thus the tailored employment of appropriate surrogates and indicator compounds within predefined boundary conditions results in a monitoring regime aimed at obtaining information that provides certainty in proper treatment performance at minimal costs.

Physicochemical properties (*e.g.*, molecular size, pK_a, hydrophobicity, volatility, and dipole moment) often determine the fate and transport of a compound during various treatment processes such as those employing high-pressure membranes.[31] Thus, the judicious selection of multiple indicators, representing a broad range of properties, enables the assessment to account for compounds that may not be currently identified ("unknowns"), as well as new compounds synthesised and entering the environment in the future (*e.g.*, new pharmaceuticals), provided they fall within the range of the properties covered by the indicators.

The underlying assumption is that absence or removal of an indicator compound during a treatment process would also ensure absence or removal of other compounds with comparable physicochemical properties. The most sensitive compounds to assess the performance of a specific treatment process will be those that are partially removed under normal operating conditions. Thus a system failure will be indicated by poor removal of the indicator compound while normal operating conditions will be indicated by partial or complete compound removal.

Predetermined changes of surrogate parameters can be used to define and assure normal operating conditions of a treatment process. Proper removal is ensured as long as the treatment process of interest is operating according to its technical specifications. It is therefore necessary to define, for each treatment process, the operating conditions under which proper removal is to be expected.[30]

Potential indicator compounds and surrogate parameters have been identified for a range of conventional and advanced water treatment processes.[30] These treatment processes have then been characterised by key removal mechanisms, such as biodegradation, chemical oxidation, photolysis, adsorption or physical separation. The potential indicator compounds were then grouped into removal categories representing "good removal" (>90%), two separate categories of "intermediate removal" (25–50% and 50–90%) and "poor removal" (<25%).

This classification of indicators into removal categories for individual treatment processes was dependant upon the physicochemical and biodegradable properties of the compounds. Whether the proposed degree of removal is achieved will depend upon the operational conditions of the treatment process (*e.g.* water matrix, oxidant contact time, membrane flux, *etc.*). Therefore, along with this classification, relevant boundary conditions were defined for each type of treatment.

Surrogate parameters are often not strongly correlated with the removal of indicator compounds occurring at nanogram per litre concentrations.[30] However, partial or complete change in the treatment performance of carefully selected surrogate parameters can indicate more general changes in the treatment performance of a unit operation or treatment train. Some surrogate parameters are also sufficiently sensitive to indicate the beginnings of performance deficiencies, which may or may not be resulting in a diminished removal of contaminants of toxicological concern. Thus, to fully assess the performance of unit operations, a combination of appropriate surrogate parameters and indicator compounds should be used.

7 Probabilistic Water Treatment Performance Assessment

Source waters supplying water recycling treatment plants tend to be variable in terms of their composition over time. That is, the diversity and concentration of dissolved chemical species changes with time. This includes short-term variability (a diurnal or weekly scale) as well as more gradual longer-term changes. Sources of this variability include local weather conditions, patterns of agricultural or industrial activities, and growth or long-term changes to the wastewater catchment.

In addition to source water variability (and sometimes because of it), unit water treatment processes also vary in their performance for reducing the concentrations of individual chemical species. Examples of the sources of treatment performance variability include variable hydraulic load, variable chemical dosing (relative to water composition), variable turbidity, as well as the gradual aging of activated carbon surfaces, filtration membranes and ultraviolet lamps.

Statistical evaluation of source water variability and treatment performance variability under normal plant operation may be achieved by summarising observed water quality using basic statistical tools associated with frequency analysis (means, standard deviations, *etc.*). Where sufficient data are available, statistical evaluation may be enhanced by fitting acquired data to distributional forms such as a normal distribution or lognormal distribution.[32] These may then be expressed as probability density functions (PDFs).

Variables such as source water concentration or treatment performance are said to be "stochastic" variables. This means that their value can not be identified as a single number since it involves a degree of variability and/or uncertainty.

Intensive sampling of the feed waters and effluents of a reverse osmosis treatment process was undertaken at one of the advanced water treatment plants supplying the WCRWP in Queensland. Monitoring data for a range of chemical species, over a period of approximately 14 months, are presented in Figure 1 (ref. 33). In this case, the data have been plotted on "lognormal probability plots". The data points are arranged from lowest to highest and plotted on a lognormal scale on the vertical axis and a probability scale on the horizontal axis. Where the data approximate a straight line, this is an indication that they fit reasonably well to a lognormal PDF. Numerical tests are also available to test the suitability of this fit.[33]

By fitting both reverse osmosis feed and reverse osmosis effluent data to PDFs, it is then possible to determine the reverse osmosis treatment performance (percentage removal of the chemical) as a stochastic variable and to describe it with a PDF.[33] This processes allows a probabilistic determination of the likelihood of the treatment process performing at a certain defined level or that a specified water quality will be achieved or exceeded.[32,34]

The variable reverse osmosis rejection performance that is apparent between the parameters shown in Figure 1 is consistent with previously reported trends. For example, excellent rejection is observed for many of the anionic and cationic species, while less significant rejection is observed for ammonia and N-nitrosodimethylamine (NDMA). Consistent with previous reports, rejection of boron at ambient pH appears insignificant.

As a stochastic variable, the final produced water quality from a water recycling scheme may be defined in terms of a PDF for each chemical contaminant of concern. The final water quality PDF will be a function of a number of other variables, most obviously source water quality to the advanced water treatment plant and the treatment performances of each of the subsequent treatment processes. Accordingly, the expected PDF for final water quality may be theoretically derived from PDFs of source water quality and treatment performance.[35,36] This concept is illustrated in Figure 2 (ref. 35).

Concentrations of a particular contaminant at each stage of the multiple barrier treatment process are represented as $C_0 \ldots C_5$ (ref. 35). The connection between the subsequent concentrations is described as a conditional PDF. For example, $F_3(C_3|C_2)$ is the probability density of the contaminant concentration following ozonation (C_3) given the concentration in the influent to the ozonation reactor (C_2). In the simplest of circumstances, the processes may be assumed to behave in a linear (first order) fashion, and the distribution may then be determined simply in terms of a ratio of effluent/influent. However, the conditional framework provides a more general approach allowing for more complex treatment process removal functions. Individual conditional PDFs may even be dependant on other descriptors of the system such as hydraulic flow rates, *etc.*

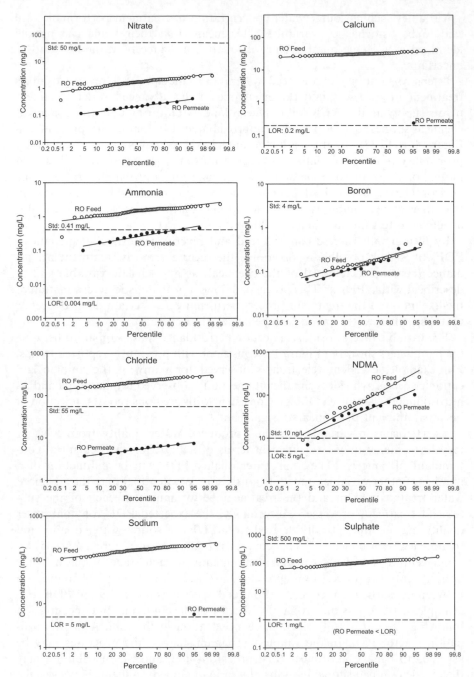

Figure 1   Concentrations of chemical constituents in reverse osmosis (RO) feed and permeate August 2008–October 2009 (ref. 33).

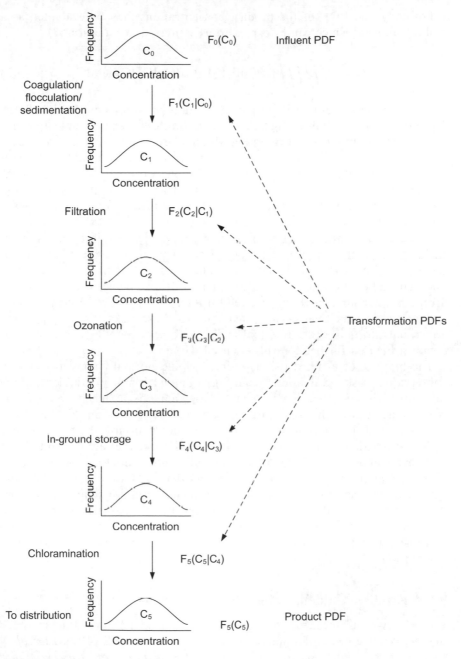

Figure 2 Conceptual diagram of multiple barrier process train and treatability distributions.[35]

Formally, the PDF of the product concentrations may be evaluated as a multiple integral which can be expressed as expressed in Equation (3):[35]

$$f_5(C_5) = \iiiint f_0(C_0)F_1F_2F_3F_4dC_0dC_1dC_2dC_3dC_4 \qquad (3)$$

Computing sums, multiplications and other transformations on multiple PDFs is a mathematically challenging task which in some cases is impossible. For this, probabilistic techniques incorporating Monte Carlo simulations can provide a powerful alternative approach.[32–34] For many modelled variables, the stochastic nature is derived from two distinct factors: variability and uncertainty. In many cases, it is desirable to separate (as much as possible) the consideration and reporting of these two factors. Such assessments are known as two-dimensional Monte Carlo methods.

Not every risk assessment requires or warrants a probabilistic assessment. Indeed, six different levels of analytical sophistication in the treatment of uncertainties in risk analysis have been identified.[37] A useful approach is to undertake a tiered evaluation whereby some risks may be deemed to be within acceptably safe limits without the need for intensive scrutiny.[37] Further efforts and resources may then be allocated only to those risks that require such additional efforts in order to establish that safe levels may be achieved or to demonstrate that further controls are needed.

The first tier of a tiered risk assessment would normally consist of a very conservative risk evaluation requiring minimal data. Variables, which may otherwise be considered stochastically, may be assigned conservative point-values based on worst case assumptions. Examples may include zero environmental degradation of a chemical in the environment or zero inactivation of a pathogen during water treatment. Hazards that are shown to be at safe levels under "tier 1 conditions" would require no further evaluation. Hazards that are not demonstrably safe under tier 1 conditions are then further subjected to a second tier assessment, which may require significantly more data and more complex analysis. A probabilistic analysis may be helpful in such circumstances since this approach involves carrying full PDFs through multiple calculations, thus avoiding the need for compounding conservative assumptions.

8   Australian Guidelines for Water Recycling

In 2008, Australia was the first country to develop national guidelines for indirect potable reuse of water with the release of Phase 2 of the *Australian Guidelines for Water Recycling (AGWR): Augmentation of Drinking Water Supplies*.[10] Phase 1 of the AGWR[38] was published in 2006 and addresses non-potable applications of reclaimed water. The Phase 2 Guidelines build upon the risk management framework described in Phase 1 and provide additional detail relating to issues specific to IPR. These guidelines provide national guidance on best practices for water recycling. They have adopted the innovative risk management framework approach pioneered in the 2004 revision of the Australian Drinking Water Guidelines.[4] These guidelines are notable for the risk management framework that they provide, rather than simply relying on end-product (recycled water) quality testing as the basis for managing water recycling schemes.

Assessment of the recycled water system must be carried out before strategies to prevent and control hazards are planned and implemented. The aim of the assessment is to provide a detailed understanding of the entire recycled water supply system, from source to end-use or receiving environment. The hazards, sources and events (including treatment failure) that can compromise recycled water quality are to be characterised and preventive measures needed to effectively control hazards and prevent adverse impacts on humans and the environment identified. Recycled water system analysis then requires the following steps, which are detailed in the guidelines:

Source of recycled water, intended uses, receiving environments and routes of exposure
- Identify source of water.
- Identify intended uses, routes of exposure, receiving environments, endpoints and effects.
- Consider inadvertent or unauthorised uses.

Recycled water system analysis
- Assemble pertinent information and document key characteristics of the recycled water system.
- Assemble a team with appropriate knowledge and expertise.
- Construct a flow diagram of the recycled water system.
- Periodically review the recycled water system analysis.

Assessment of water quality data
- Assemble historical data about sewage, greywater or stormwater quality, as well as data from treatment plants and of recycled water supplied to users; identify gaps and assess reliability of data.
- Assess data (using tools such as control charts and trends analysis) to identify trends and potential problems.

Hazard identification and risk assessment
- Define the approach to hazard identification and risk assessment, considering both public and ecological health.
- Periodically review and update the hazard identification and risk assessment to incorporate any changes.
- Identify and document hazards and hazardous events for each component of the recycled water system.
- Estimate the level of risk for each identified hazard or hazardous event.
- Consider inadvertent and unauthorised use or discharge.
- Determine significant risks and document priorities for risk management.
- Evaluate the major sources of uncertainty associated with each hazard and hazardous event and consider actions to reduce uncertainty

Examples of potentially hazardous events that could be considered are provided in the guidelines. The level of risk for each hazardous event can be estimated by identifying the likelihood that it will happen and the severity of the consequences if it does. Guidelines and criteria developed for specific combinations of source water and end-use should be referred to when estimating risk. The likelihood and consequences can then be combined to provide a qualitative estimation of risk, using a suitable risk matrix. Risks that are judged to be very high will generally be the focus of critical control points.

Health-related guideline values for chemicals are provided in the Phase 2 guidelines. Where possible, these guideline values were acquired from existing guidelines and standards, with the Australian Drinking Water Guidelines being the primary source. Where no existing guideline values could be identified, guidelines were developed from available health, toxicological or structural information.

Guidelines for dioxins, furans and polychlorinated biphenyls (PCBs) were developed using recommended tolerable intakes developed by the NHMRC accounting for the combined total exposure for multiple chemicals. Using toxicity equivalency factors as evaluated by the World Health Organisation,[39] a total concentration guideline value of $0.016\,\mathrm{ng\,l^{-1}}$ toxicity equivalents (TEQs) was determined for this group of chemicals.

Guideline values for human pharmaceuticals were derived from lowest daily therapeutic doses divided by uncertainty factors of 1000–10 000. These uncertainty factors were derived in the same way as those described in Section 3, with an addition factor of 10 for cytotoxic drugs and steroidal hormones. Guidelines for pharmaceuticals used for agricultural or veterinary purposes were developed from acceptable daily intake (ADI) values established by a range of international food and health agencies.

Where neither existing guidelines, nor relevant toxicological data for developing guidelines were available, a quantitative structure–activity relationship approach was used as the method for determining thresholds of toxicological concern (TTCs).

The guidelines note that an extensive range of parameters can be used to represent a risk. They acknowledge that is not physically or economically feasible to test for all parameters, nor is it necessary. The list of guideline values determined and provided is not intended to be regarded as a mandatory set of parameters to be included in monitoring programs. However, key characteristics that must be considered for system performance verification include:

- Microbial indicator organisms.
- Health-related chemicals, including:
  - Those identified in the ADWG;
  - Key organic chemicals of concern (*e.g.* NDMA); and
  - Indicators or index chemicals for organic chemicals (*e.g.* contraceptive hormones).
- Biological activity.

The choice of specific parameters must be informed by hazard identification and risk assessment. These, in turn, should be informed by consideration of source water quality, potential agricultural and industrial inputs, treatment processes, chemicals and by-products, and receiving water quality.

9 Conclusions

The careful and effective management of chemical contaminants in IPR schemes requires a variety of relatively sophisticated techniques for assessing and identifying key risks to human health. The range of chemicals present in source waters supplying IPR schemes is diverse and far from fully characterised. Nonetheless, established toxicology of some individual chemical species can be used in conjunction with chemical analysis to determine safe drinking water concentrations for some substances. However, this approach has significant limitations and, for a more complete characterisation of water safety, should be supplemented with direct toxicity testing techniques. Unfortunately, *in vivo* toxicity testing suffers from practical limitations including complex experimental design, high costs, long testing periods and animal welfare considerations. While *in vitro* methods are able to overcome some of these limitations, there remains a high degree of uncertainty regarding the interpretation and relevance to human health for many assays.

The use of surrogate and indicator species to monitor water treatment performance is highly promising, but there are few examples of effective implementation other than bulk parameters such as conductivity and total organic carbon. These bulk parameters do not necessarily capture the more subtle effects that may be relevant for some key trace chemical contaminates of concern. Probabilistic water treatment performance provides an approach for defining water treatment performance of individual unit processes and to theoretically combine those processes in order to determine probable final water qualities, including concentrations of chemical species below available analytical detection limits.

All of these approaches to chemical water quality assessment offer a diverse range of advantages, but they all come with significant limitations. Accordingly, the most prudent approach to comprehensive management of chemicals in IPR schemes appears to be the adoption of a range of these techniques suitable for the characteristics of the scheme being assessed and the degree of risk management that is to be imposed. Experience with existing IPR schemes informs us that the careful selection and application of chemical risk assessment and risk management practices can effectively enhance the provision of a safe and sustainable IPR water supply.

References

1. National Research Council, *Risk Assessment in the Federal Government: Managing the Process,* National Academy Press, Washington, DC, USA, 1983.

 2. L. Ritter, C. Totman, K. Krishnan, R. Carrier, A. Vezina and V. Morisset, *J. Toxicol. Environ. Health. B.*, 2007, **10**, 527–557.
 3. US Environmental Protection Agency, *A Review of the Reference Dose and Reference Concentration Processes, Risk Assessment Forum*, EPA/630/P-02/002F, Washington, DC, USA, 2002.
 4. National Water Quality Management Strategy, *Australian Drinking Water Guidelines*, National Health and Medical Research Council, Natural Resource Management Ministerial Council, Government of Australia, Canberra, Australia, 2004.
 5. World Health Organization, *Guidelines for Drinking-Water Quality*, 3rd edn, Geneva, Switzerland, 2004.
 6. US Environmental Protection Agency, *Risk Assessment Guidance for Superfund, Volume 1: Human Health Evaluation Manual (Part A)*, 1989.
 7. R. A. Howd and A. M. Fan, *Risk Assessment for Chemicals in Drinking Water*, John Wiley & Sons, Inc., Hoboken, New Jersey, USA, 2008.
 8. B. W. Schwab, E. P. Hayes, J. M. Fiori, F. J. Mastrocco, N. M. Roden, D. Cragin, R. D. Meyerhoff, V. J. D'Aco and P. D. Anderson, *Regul. Toxicol. Pharmacol.*, 2005, **42**, 296–312.
 9. S. A. Snyder, *Ozone: Sci, Eng.*, 2008, **30**, 65–69.
10. Natural Resource Management Ministerial Council, Environment Protection and Heritage Council and National Health and Medical Research Council, *Australian Guidelines for Water Recycling: Managing Health & Environmental Risks (Phase 2) - Augmentation of Drinking Water Supplies*, Canberra, Australia, 2008.
11. R. Kroes, J. Kleiner and A. Renwick, *Toxicol. Sci.*, 2005, **86**, 226–230.
12. I. C. Munro, A. G. Renwick and B. Danielewska-Nikiel, *Toxicol. Lett.*, 2008, **180**, 151–156.
13. C. Rodriguez, A. Cook, P. Van Buynder, B. Devine and P. Weinstein, *Water Sci. Technol.*, 2007, **56**, 35–42.
14. C. Rodriguez, P. Weinstein, A. Cook, B. Devine and P. Van Buynder, *J. Toxicol. Environ. Health A*, 2007, **70**, 1654–1663.
15. A. W. Olivieri, D. M. Eisenberg, R. C. Cooper, G. Tchobanoglous and P. Gagliardo, *Water Sci. Technol.*, 1996, **33**, 285–296.
16. National Research Council, *Issues in Potable Reuse: The Viability of Augmenting Drinking Water Supplies with Reclaimed Water,* National Academy Press, Washington, DC, USA, 1998.
17. L. W. Condie, W. C. Lauer, G. W. Wolfe, E. T. Czeh and J. M. Burns, *Food Chem. Toxicol.*, 1994, **32**, 1021–1030.
18. A. de Peyster, R. Donohoe, D. J. Slymen, J. R. Froines, A. W. Olivieri and D. M. Eisenberg, *J. Toxicol. Environ. Health*, 1993, **39**, 121–142.
19. NEWater Expert Panel, Singapore Water Reclamation Study: Expert Panel Review and Findings, 2002.
20. D. Schlenk, D. E. Hinton and G. Woodside, *Online Methods for Evaluating the Safety of Reclaimed Water,* Water Environment Research Foundation (WERF), Alexandria, VA, USA, 2007.

21. T. Asano and J. A. Cotruvo, *Water Res.*, 2004, **38**, 1941–1951.
22. M. Balls, A. M. Goldberg, J. H. Fentem, C. L. Broadhead, R. L. Burch, M. F. W. Resting, J. M. Frazier, C. F. M. Hendriksen, M. Jennings and M. D. O. van der Kamp, *ATLA. Alternatives Lab. Animals*, 1995, **23**, 838–866.
23. F. D. L. Leusch, S. J. Khan and H. Chapman, in *Water Reuse and Recycling*, ed. S. J. Khan, R. M. Stuetz and J. M. Anderson, UNSW Press, Sydney, NSW, Australia, 2007, pp. 296–303.
24. M. H. Nellor, R. B. Baird and J. R. Smyth, *Health Effects Study Final Report*, County Sanitation Districts of Los Angeles County, Whittier, CA, USA, 1984.
25. J. M. Montgomery, *Operation, Maintenance and Performance Evaluation of the Potomac Estuary Experimental Water Treatment Plant*, Montgomery (James M.) Consulting Engineers Inc. Pasadena, CA, Alexandria VA, USA, 1983.
26. Western Consortium for Public Health, *The City of San Diego - Total Resource Recovery Project*, San Diego, CA, USA, 1992.
27. K. Thompson, R. C. Cooper, A. W. Olivieri, D. Eisenberg, L. A. Pettegrew and R. E. Danielson, *Desalination*, 1992, **88**, 201–214.
28. CH2M Hill, *Tampa Water Resource Recovery Project: Pilot Studies*, Tampa, FL, USA, 1993.
29. J. Hemmer, C. Hamann and D. Pickard, *Water Reuse Symposium, Dallas TX, Denver CO, USA*, 1994.
30. J. E. Drewes, D. L. Sedlak, S. Snyder and E. Dickenson, *Development of Indicators and Surrogates for Chemical Contaminant Removal during Wastewater Treatment and Reclamation - Final Report*, WateReuse Foundation, Alexandria, VA, USA, 2008.
31. C. Bellona, J. E. Drewes, P. Xu and G. Amy, *Water Res.*, 2004, **38**, 2795–2809.
32. D. Eisenberg, J. Soller, R. Sakaji and A. Olivieri, *Water Sci. Technol.*, 2001, **43**, 91 99.
33. S. J. Khan, *Quantitative Chemical Exposure Assessment for Water Recycling Schemes*, National Water Commission Waterlines Report Series No. 27, Canberra, Australia, 2010.
34. S. J. Khan and J. A. McDonald, *Water Sci. Technol.*, 2010, **61**, 77–83.
35. C. N. Haas and R. R. Trussell, *Water Sci. Technol.*, 1998, **38**, 1–8.
36. H. Tanaka, T. Asano, E. D. Schroeder and G. Tchobanoglous, *Water Environ. Res.*, 1998, **70**, 39–51.
37. M. E. Paté-Cornell, *Reliability Eng. Syst. Safety*, 1996, **54**, 95–111.
38. National Resource Management Ministerial Council and Environment Protection & Heritage Council, *Australian Guidelines for Water Recycling: Managing Health & Environmental Risks (Phase 1)*, 2006.
39. M. Van den Berg, L. S. Birnbaum, M. Denison, M. De Vito, W. Farland, M. Feeley, H. Fiedler, H. Hakansson, A. Hanberg, L. Haws, M. Rose, S. Safe, D. Schrenk, C. Tohyama, A. Tritscher, J. Tuomisto, M. Tysklind, N. Walker and R. E. Peterson, *Toxicol. Sci.*, 2006, **93**, 223–241.

Nanotechnology for Sustainable Water Treatment

MATT HOTZE* AND GREG LOWRY

ABSTRACT

As world water demand continues to grow there is a critical need to develop sustainable water treatment solutions. This chapter describes the potential for nanomaterials to improve the sustainability of water treatment. Nanomaterial-driven advances in disinfection, oxidation, membrane separation and groundwater remediation are discussed with a view towards their potential to improve existing technologies. Disinfection technologies include oligodynamic processes with silver nanoparticles to effectively inactivate microorganisms without disinfection byproducts being formed. Oxidation technologies include metal oxide semiconductors and fullerene-based sensitisers acting as light-driven catalysts. Membrane separation processes include the embedding of materials such as zeolites, carbon nanotubes and metal oxides to improve selectivity and reduce fouling. Remediation technologies include iron particles designed to target and transform waste compounds *in situ*. These and other emerging water treatment technologies must be assessed with life-cycle analysis to determine the full materials and embodied energy costs of acquiring raw materials, manufacturing, use and end of life for the materials contained within each process. These costs must be weighed against the potential benefits for water treatment to determine their sustainability.

---

*Corresponding author.

Issues in Environmental Science and Technology, 31
Sustainable Water
Edited by R.E. Hester and R.M. Harrison
© Royal Society of Chemistry 2011
Published by the Royal Society of Chemistry, www.rsc.org

1 Introduction

Improving water treatment is an important issue in achieving sustainability. Within the next century the world population is expected to increase from *ca.* 6 billion to *ca.*9.5 billion people. Estimates vary, but nearly 2 billion people currently suffer from water scarcity (*i.e.*, $<1000\,m^3$ water per year). The possibility exists that 1 billion or more water-scarce people will be added the world population by 2025. The World Health Organisation (WHO) reports that 90% of water goes untreated after use in underdeveloped countries,[1] thereby degrading the quality of surface waters in those regions. Moreover, 15% of the world's population lacks access to improved water supplies and 42% of the population lacks adequate sanitation (see Figure 1). These facts relate directly to health issues in downstream communities.

Water treatment is chemical and energy intensive and world energy consumption *per capita* continues to grow for the foreseeable future. Correspondingly, there is an increase in toxic chemicals released into the environment each year and these already are on the magnitude of billions of kilograms. Finally, the American Society of Civil Engineers (ASCE) estimates that 11 billion dollars is needed to replace drinking water infrastructure and 390 billion dollars needs to be invested in wastewater treatment in the United States alone.[2] These are tremendous challenges for meeting sustainability targets. Improving water treatment technologies is an important step toward sustainability.

Nanomaterials have a potential to improve the sustainability of water treatment processes by increasing water supply and quality as well as reducing energy usage. Nanomaterials have a two-part definition. Firstly, the size scale of the material – either as a particle or a surface feature – lies between 1 and 100 nm (ref. 3). Secondly, the properties of that material must be different from those of material with the same chemical composition at larger scales (*i.e.* a true "nano" effect must be exhibited).[4] The second part of this definition is most critical to understanding why nanomaterials could improve water treatment

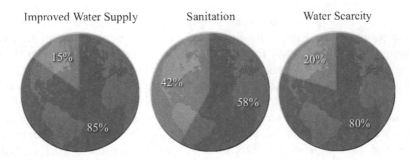

Figure 1   Approximately 15% of the world's population does not have access to an improved water supply; *ca.* 42% of the world's population does not have access to adequate sanitation; and *ca.* 20% of the world's population lives in water-scarce conditions (*i.e.* $<1000\,m^3$ per year).

processes. The unique surface features and reactivity of nanomaterials when compared with larger sized bulk materials potentially leads to lower raw material needs, higher reaction rates and surface chemistry specificity. This chapter examines some water treatment processes for disinfection, membrane separation and groundwater remediation that could improve upon existing technologies by harnessing these improvements.

2   Disinfection and Oxidation Technologies

Disinfection is usually performed as a polishing step during water treatment to inactivate remaining microbes (*e.g.* viruses and bacteria) and to provide residual oxidant throughout a drinking-water distribution system. Chlorination is still the most common disinfection method employed. However, the use of chlorine is well known to cause disinfection byproducts that induce chronic effects in humans.[5] In response to this, alternative disinfection processes have been developed, such as chlorine dioxide, chloramines, ozonation and ultra-violet irradiation. But these processes come with their own potential drawbacks, such as higher cost, lack of a residual and potential formation of new types of disinfection byproducts.[6] The aims of disinfection nanotechnologies, therefore, are to reduce cost and avoid formation of disinfections byproducts. Two significant veins of research in this area are metal nanoparticles with an oligodynamic effect and photo-driven processes based on metal oxides or carbon nanomaterials.

2.1   Oligodynamic Processes

Very low concentrations of metals such as cobalt, copper, nickel, silver, titanium and zinc can be bactericidal and virucidal *via* the oligodynamic effect.[7] While there are a variety of metals that exhibit this effect, silver has the most widely effective potency and accordingly attracts the most attention for applications in water treatment. Silver ions have long been known to be bactericidal.[8] The source of their bactericidal properties is still not clearly agreed upon in literature. However, some strong hypotheses have been made that silver ions affect bacteria by inactivating the thiol groups on critical enzymes, reducing the ability of DNA to replicate within the cell and causing structural changes to the cell membrane.

More recently, silver nanoparticles (AgNPs) have also been shown to deleteriously affect bacteria *via* several mechanisms: nanoparticle attachment to the cell membrane surface, altering permeability and respiration; particle penetration inside the cell and subsequent interaction with compounds containing sulfur and phosphorus (*e.g.* DNA); and the dissolution of the particles, releasing silver ions that also affect the bacteria (see Figure 2; ref. 8). AgNP size may also be related to the magnitude of the effects on bacteria.[9]

AgNPs have been suspended on several substrates including granular activated carbon, activated carbon fibres, polyurethane, zeolites and ceramics, and

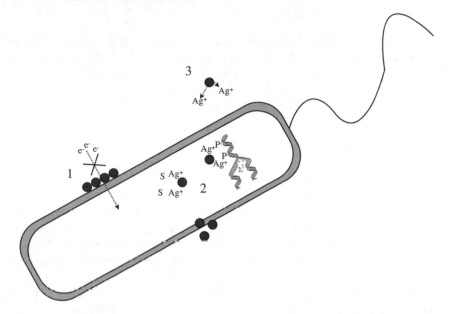

Figure 2 Hypothesised mechanisms of silver nanoparticle inactivation of bacteria. 1: Blocking cell respiration at the wall. 2: Interaction with sulfur- and phosphorus-containing compounds (*e.g.* DNA) within the cell. 3: Release of silver ions from the particles.

have displayed effective inactivation of pathogens in water.[7] Silver particles are typically formed on substrates by reduction of silver salts (*e.g.* AgNO₃) either thermally or chemically.[10] The reduction of Ag^+ to Ag^0 results in nucleation of particles on the substrate (*e.g.* the pores within a stem of a rice paper plant).[10] Alternative synthesis routes of AgNPs that eliminate the use of environmentally problematic reducing agents (*i.e.* "green chemistry") are currently being investigated.[11]

Early research indicates significant potential for AgNPs for point of use (POU) disinfection. AgNP-coated polyurethane foams have been shown to serve effectively as POU filters by removing up to $10^6 \, CFU \, ml^{-1}$ at a flow rate of $0.51 \, min^{-1}$ (ref. 12). Additionally, both carbon and glass fibres have been impregnated with AgNPs and high disinfections achieved in both cases with a variety of organisms.[13,14] In another case, POU ceramic water filters (fabricated in a sustainable manner using local soils in Mexico and Guatemala) were coated with silver nanoparticles to effectively improve bacteria removal in effluent drinking waters.[15] However, in a different study of AgNP-coated POU filters, 3.8–4.5 log reduction in bacteria colony forming units (CFUs) was achieved initially, but dropped to 0.2–2.5 log reduction after exposure to a high load of bacteria (*ca.*$10^6 \, CFU \, ml^{-1}$). Reapplication of silver did not result in the recovery of the filter to original disinfection levels.[16] Thus, though promising, more research is needed to optimise the performance of filters modified with nanoparticles to improve water treatment.

2.2   Photo-Driven Processes

In nanomaterial-based disinfection work, two types of photo-driven reactive oxygen species (ROS) producing processes exist: photocatalysis with metal oxides and photosensitisation with carbon-based materials. Collectively, materials that utilise oxidation to disinfect or treat other contaminants in water are known as advanced oxidation processes (AOPs).

2.2.1   Photocatalytic Semiconductors

Titanium dioxide (TiO_2) is the most promising and most studied material proposed as a photocatalyst for AOPs. TiO_2 acts as a semiconductor photocatalyst when light of energy equal to or greater than its intrinsic band gap (3.0–3.2 eV or wavelength, $\lambda \leq 390$ nm) is absorbed by the material. Electrons excited to the conduction band (CB) can then either migrate to the surface, reducing oxygen to form superoxide ($O_2{}^{\cdot-}$), or recombine with valence band (VB) holes. Recombination results only in the generation of heat. If VB holes are not recombined they migrate to the surface and oxidise water to form hydroxyl radicals ($OH^{\cdot}$) (see Figure 3). Semiconductor photocatalysts with different band gaps are listed in Table 1.

While these semiconducting materials have been known for some time,[17] the advent of nanotechnology promises to overcome some of the challenges that hinder their practical application. These include enhancement of photocatalytic activity, controllability of structural properties, immobilisation of the material

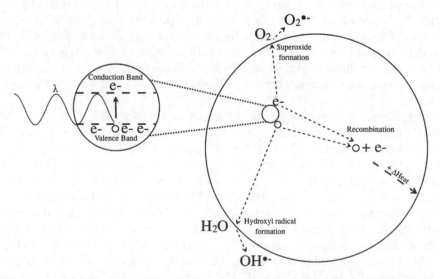

Figure 3   Semiconductor photocatalysis forms free electrons in the conduction band (CB) and holes are left behind in the valence band (VB). Electrons and holes can migrate to the surface to reduce oxygen and form superoxide or oxidise water to form hydroxyl radicals. Recombination can also occur within the crystal structure, releasing heat.

Table 1 Examples of nanoparticle photocatalysts and their corresponding band gap energy.

Photocatalyst material	Band gap (eV)
C_{60}	2.3
CdS	2.4
α-Fe_2O_3	3.1
Fe_2O_3	2.2
TiO_2 (anatase)	3.2
TiO_2 (rutile)	3.0
WO_3	2.7
ZnO	3.2

on surfaces and narrowing the band gap energy to increase quantum yields from the solar spectrum.[18] Sol-gel processes are used to synthesise TiO_2 particles and, because of the analytical tools developed in parallel with nanomaterials, the changes in structure resulting from varying these synthesis procedures can now be controlled readily.[19] For example, a sol-gel synthesis in combination with surfactant allows for the creation of porous material with very high surface area ($150\,m^2\,g^{-1}$) and improved catalytic activity.[20] UV irradiation is the only means to create reactive oxygen species (ROS) in solution (*e.g.* $O_2^{\bullet-}$ or $OH^{\bullet}$) because jumping the band gap of TiO_2 requires wavelengths of *ca.* 400 nm or lower. Doping TiO_2 crystals with other elements (*e.g.* nitrogen) that decrease the band gap allows for the utilisation of visible light.[21,22] However, doping also causes a problem by promoting the undesirable recombination of electron and hole pairs, but the addition of metals and metal oxides to the synthesis to form hybrid semiconductor nanomaterials (HSN)[23] reduces this effect.[24,25] Improving the oxidation properties of TiO_2 through nanotechnology is very promising for improving water treatment. HSNs have demonstrated early success. For example, TiO_xN_y/PdO HSNs were tested under visible light and antimicrobial activity was enhanced above that of materials that were only doped with nitrogen or coated with PdO (ref. 26); bacterial spores, which were normally extremely resistant to treatment, were inactivated at high rates.[27]

2.2.2 Fullerene Photosensitisation

Fullerenes are another type of photoactive material that is of interest for AOPs in water treatment. These behave as photosensitisers, generating ROS such as superoxide ($O_2^{\bullet-}$) and singlet oxygen ($^1O_2$) *via* type I and type II photosensitisation pathways, given appropriate suspension conditions (see Figure 4). Electrons within the molecular orbitals of C_{60} can be promoted by light into the singlet state, $^1C_{60}$; this is an unstable state that rapidly and efficiently transforms into the triplet state, $^3C_{60}$, *via* a process known as intersystem crossing.[28] Once in the triplet state a variety of reactions can occur: Type I reduction to form radical fullerene ($C_{60}^{\bullet-}$) and subsequent formation of $O_2^{\bullet-}$; Type II

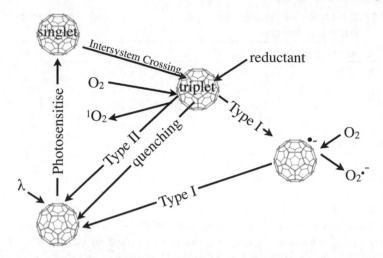

Figure 4   Fullerene photochemistry. (Adapted with permission from the American
Chemical Society).[28]

sensitisation of oxygen to form $^1O_2$; quenching reactions such as triplet-triplet
annihilation and self-quenching that result in no ROS formation, see Equations
(1)–(4).

$$^3C_{60} + O_2 \rightarrow C_{60} + {}^1O_2 \quad (\textit{type II reaction}) \tag{1}$$

$$^3C_{60}^{\bullet -} + O_2 \rightarrow C_{60} + O_2^{\bullet -} \quad (\textit{type I reaction}) \tag{2}$$

$$^3C_{60} + C_{60} \rightarrow 2\,C_{60} \quad (\textit{self-quenching}) \tag{3}$$

$$^3C_{60} + {}^3C_{60} \rightarrow C_{60} + {}^3C_{60} \quad (\textit{triplet-triplet annihilation}) \tag{4}$$

Quenching processes are promoted when unmodified cages aggregate; however,
when fullerenes are functionalised (*e.g.* fullerol), the cages are able to produce
ROS in aqueous suspension.[28–30]

Researchers have been able to utilise fullerene ROS production to demonstrate significant inactivation of bacteriophages (viruses) in aqueous suspension
when compared with UV light alone.[31,32] In related work, fullerenes with
anionic, neutral and cationic functional groups have been produced to test their
efficacies as photoactivated bactericides (see Table 2; ref. 33). Cationic functionalised fullerenes were shown to be the most effective disinfectant for bacteria and viruses because they were the most efficient producers of ROS, and
electrostatics allowed for the close association of microbes (negatively charged)
and fullerene cages (positively charged).

Table 2 Fullerenes functionalised with anionic, neutral, and cationic groups for inactivation of bacteria and viruses. (Adapted with kind permission from the American Chemical Society).[33]

Chemical Structure	R'	R''	Charge at pH 7
(R' R'')6	-CO₂H	-CO₂H	anionic
	-CO₂H	O=⟨⟩-N⟨O⟩	anionic
	-NHCH(CH₂OH)₂	-NHCH(CH₂OH)₂	neutral
	-CO₂(CH₂)₂NH₃⁺CF₃CO₂⁻	-CO₂(CH₂)₂NH₃⁺CF₃CO₂⁻	cationic

3 Nanotechnology Improving Membranes for Water Treatment

The use of membranes in water processing is a mature technology. The three remaining challenges to improve membrane separation processes to make them sustainable have been identified as: increasing water permeability to reduce energy cost of clean water; improved control of permeate selectivity in membrane design; and surfaces that prevent fouling.[34,35] The ability to manipulate structure, surface chemistry and reactivity at the nanoscale has led researchers to attack these problems with new strategies.

3.1 Nanocomposite Membranes

The concept of nanocomposite membranes has been proposed for nearly 20 years.[36] However, research on this topic has not developed until recently. The main thrusts of this work include: thin film with zeolite sieve pathways embedded;[37,38] membranes embedded with TiO_2 for photoactive surfaces;[39] silver NPs added to membranes for biocidal purposes;[40–42] and carbon nanotube-embedded membranes.[43]

Zeolites, minerals with over 175 unique frameworks combining elements such as sodium, potassium, calcium and magnesium into various aluminosilicate crystal structures, can be embedded in thin film polyamide membranes for the purpose of creating preferential "molecular sieve nanoparticle" pathways in reverse osmosis membranes.[37] These sieves would have the surface properties of aluminium-rich zeolite (superhydrophilic and negatively charged) while providing preferential flow paths and maintaining rejection by steric and Donnan (charge) exclusion forces (see Figure 5). Recently, it has been

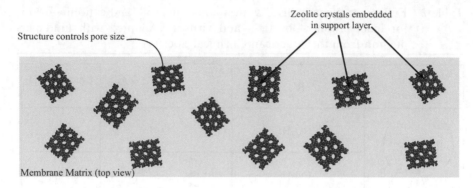

Figure 5   A top-down idealised depiction of a membrane embedded with zeolite materials. Control of zeolite structure can sieve small molecules through membranes selectively to improve flux and rejection rates.

demonstrated that the size of the nanoparticle zeolite can play a role in determining the membrane properties. Smaller zeolites (diameter *ca.*100 nm) provide higher flux as well as higher solute rejection, while larger zeolites (diameter *ca.*300 nm) provide the most favourable surface potential properties.[38]

Silver nanoparticle (AgNP) embedded membranes are being investigated for water treatment. These membranes have AgNPs embedded to prevent bio-fouling on the surface of the membranes (a major challenge of utilising membranes in water treatment processes). Section 2.1 discussed some of the mechanisms of silver toxicity that have been explored (see Figure 2). Silver nanoparticles can either be embedded in polymer nanocomposites[40-42] or ceramic composites.[44] The nanoparticles can also be formed *ex situ* prior to incorporating them into the membrane,[40] or *in situ* by reactions with silver ions present in the formed membrane.[42] *Ex situ* formation allows for more control of silver particle size and morphology characteristics, and *in situ* formation allows for more control over homogeneous distribution of the particles within the membrane.[41] Silver nanocomposites in this work demonstrated a potential for application as macroporous membrane spacers to inhibit biofilm growth on downstream membrane surfaces.

TiO$_2$-impregnated membranes are another class of nanoreactive membranes that are being currently investigated for water treatment. TiO$_2$ membranes can either be polymeric (*e.g.* TiO$_2$ particles embedded into a polymer matrix)[39] or ceramic (*e.g.* TiO$_2$ films formed on the surface of a ceramic[45] or TiO$_2$ "nano-wires" forming a microporous membrane).[46] The concept behind these membranes is to utilise the radicals (*e.g.* O$_2$·$^-$ or OH·) produced by photoactive TiO$_2$ to reduce membrane fouling (see section 2.2.1). Many studies have successfully demonstrated that the TiO$_2$ semiconductor does remain active enough to degrade organic compounds such as methylene blue[45] and prevent membrane fouling in the case of TiO$_2$ 'nanowire' membranes.[46]

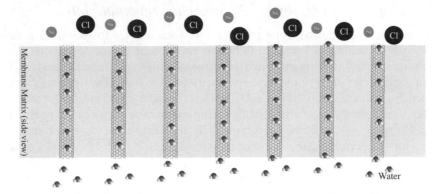

Figure 6 Idealised depiction of selectivity for a membrane embedded with nanotube pores. Water molecules face very low friction within nanotube pore structures.

3.2 Nanotube Embedded Membranes

Nanotube embedded membranes can theoretically be formed by both carbon nanotubes (CNT)[43] and inorganic nanotubes such as boron nitride and silicon.[47] Controlling tube sizes allows membrane selectivity to be precisely tuned.[43] Figure 6 illustrates this concept with water molecules passing through the pores formed by the tubes, and salt components such as sodium and chloride ions being rejected. In the case of carbon nanotubes molecular dynamics (MD) simulations have demonstrated that the smooth walls of the tubes allow for water fluxes that are four to five orders of magnitude higher than predicted from fluid-flow theory, resulting from a "nearly frictionless" interface with the CNT wall.[48] In practice, the most common means to form these vertically aligned membranes is by catalytic chemical vapour deposition, allowing for layer-by-layer formation of CNTs on a catalyst surface.[49] Although some experiments have been performed with these membranes, up to this point studies have utilised monodisperse suspensions of particles to check pore sizes. Therefore, little beyond MD simulation is known about selectivity in a polydisperse/complex water suspension. Much of the focus remains on the preparation technique of the vertically aligned CNTs (*e.g.* increasing the number of CNT pores per unit membrane square area).[50] A more expansive review on the use of carbon nanotubes in membranes is provided by Mauter and Elimelech.[51]

Tubes formed from boron nitride (BN) have also been investigated with MD simulation. BN tubes may prove to be superior for membrane applications because of improved chemical stability[52] and water permeation[53] as compared with CNTs. MD simulation indicates that 100% sodium chloride rejection could theoretically be achieved from concentrations up to 1 M utilising BN tubes.[52] However, there has been no experimental evidence gathered with BN tubes so it remains unclear if these simulations will translate into practical application.

3.3   Monitoring Membrane Failure with Nanomaterials

Conventional polymeric membranes can fail when utilised for water treatment processes. Detection of breakage in membranes can be performed online using methods such as turbidity and particle counters. However, these methods suffer from lack of sensitivity and non-specificity as they count all particle types. Recently, magnetic iron NPs (Fe_3O_4) with a diameter of 35 nm in the influent water were tested as a method of detecting membrane breaks in a polymeric ultrafiltration hollow fibre membrane.[54] By measuring the magnetic susceptibility of the permeate, the researchers were able to detect 1% breakage rate down to an iron NP concentration of 1.2 ppm while the unit was online. Particles are on the same size scale as viruses (*ca.* 20 nm); therefore, breakages allowing viruses through would also allow iron NP permeation. This is a functional implementation of nanomaterials with existing treatment processes to achieve reductions in downtime and energy costs with membrane processes.

4   Groundwater Remediation Using Nanotechnology

Fresh groundwater represents less than one percent of the total water on earth, but is a significant water source (*e.g.* 53% of drinking water sources in the United States).[55] Groundwater contaminants include arsenic, chromium (VI), dioxins, mercury, perchlorate, tetrachloroethylene (PCE) and trichloroethylene (TCE) amongst others. Perchlorate – which is responsible for contaminating water in thirty five US states – is a highly soluble component of rocket fuel that blocks uptake of iodine by the thyroid gland. Therefore the United States Environmental Protection Agency (USEPA) has set the safe drinking water dose at 0.7 µg per kg of body weight per day.[56] Chlorinated solvents such as trichloroethylene (TCE) and tetrachloroethylene (PCE) have been used since the mid-1920s in a wide variety of industries, especially in dry-cleaning and metal-degreasing processes. Improper storage, disposal, transit and handling of these solvents lead to significant groundwater and soil contamination.[57] Due to carcinogenicity and toxicity, the maximum contamination level (MCL) of TCE and PCE in drinking water is regulated by USEPA to be 5 ppb, while the maximum contamination level goal (MCLG) is zero.[58]

Two general types of processes can be utilised to remediate contaminated groundwater sites: extraction or transformation. Extraction removes the polluting chemical from the groundwater system, and methods include excavation as well as pump and treat. However, excavation is not practical for most cases and pump and treat systems are energy intensive.[59] Transformation either mineralises the compound or converts it into less toxic forms. Methods for transformation include chemical oxidation and enhanced bioremediation.[60] Three nanotechnologies for groundwater remediation are: electrically switched ion exchange (ESIX); bimetallic nanoparticles; and reactive iron nanoparticles (RNIP). The first is an example of an ion exchange technology for pump and treat applications and the latter two improve chemical oxidation transformation processes in the subsurface.[61,62]

4.1 Electrically Switched Ion Exchange (ESIX)

Electrically switched ion exchange (ESIX) is a process that uses conductive ion-exchange materials to reversibly adsorb ions from a waste stream. For groundwater remediation, the waste stream would be pumped to the surface and treated, as is the case with typical ion-exchange processes. ESIX is reversible because conductive ion-exchange materials can have an anodic or cathodic charge applied.[63] To remove waste cationic species (M^+) an anodic potential (V) is applied oxidising the adsorption site (X^-), see Equation (5).

$$e^- + M^+ + X \xleftrightarrow{V} |M^+X^-| \tag{5}$$

When the process needs to be reversed a cathodic potential is applied to the material and the waste ion is released. The same principle applies for removal of anionic waste ions, see Equation (6).

$$M^- + X \xleftrightarrow{V} e^- + |X^+M^-| \tag{6}$$

Perchlorates are typically treated in anion-exchange resin columns where the anion waste (ClO_4^-) is removed. Periodically the resins need to be regenerated in order for acceptable removal efficiencies to be maintained. However, the regeneration wastes contain perchlorate and become secondary wastes.[64] In the case of ESIX, perchlorates can be removed by a conductive polypyrrole (pPy) deposited on a film of conductive and high surface area carbon nanotubes (see Figure 7; ref. 61).

4.2 Nano Reactive Zero Valent Iron Particles for in situ Groundwater Remediation

Nano reactive zero valent iron (NZVI) particles are used to reduce and transform or immobilise contaminants in groundwater.[65–67] If these nanoparticles are utilised at full scale, the challenge is to effectively bring the NZVI surface into contact with pollutants. Loss of reactivity, low particle mobility and non-specificity of particle delivery are three significant obstacles. Physicochemical properties that most significantly affect particle reactivity are particle crystallinity and chemical composition.[65,67–69] Because of the large mass of material required for remediation, treatment costs depend on particle reactivity, reactive lifetime and particle efficiency. Bare NZVI particles will rapidly oxidise, forming a non-reactive iron oxide shell; this can reduce reactivity because electrons cannot easily be delivered to the surface for reduction reactions (see Figure 8).

Several methods of synthesis have been attempted to increase the reactivity of NZVI. Table 3 lists the properties of the four types of iron nanoparticles examined here, including: reactive nanoscale iron particles (RNIP) available commercially from TODA (Koygo, Japan); iron particles synthesised from

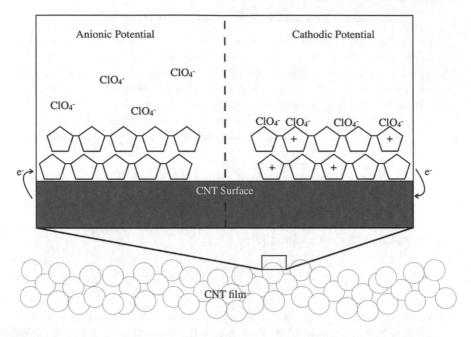

Figure 7 ESIX for perchlorate removal. Cathodic potential is applied to the CNT surface to oxidise polypyrrole (pPy) deposited on the surface and perchlorate ions (ClO_4^-) are removed from solution. Applying an anionic potential releases the perchlorate ions.

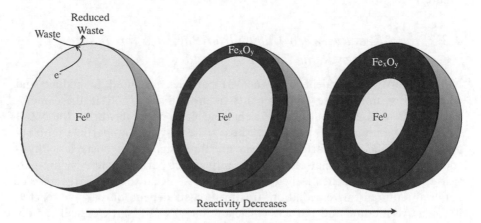

Figure 8 RNIP is oxidised over time and reactivity decreases.

sodium borohydride reduction of ferrous iron (Fe^B; ref. 65); iron particles formed by sputtering iron atoms (Fe^{SP}; ref. 70); and bimetallic particles (Pd/Fe^B; ref. 68; see section 4.3). Evidence gathered from transmission electron microscopy (TEM), energy dispersive spectroscopy (EDS), electron energy-loss

Table 3 Characteristics of NZVI synthesised by different methods. [65,68,70]

Type	Source	Particle size (nm)	Surface area ($m^2 g^{-1}$)	Crystallinity	Major phase	Minor phase
RNIP	Commercial	40 to 60	23	Crystalline	α-Fe^0	Magnetite
Fe^B (ref. 65)	Laboratory	30 to 40	36.5	Amorphous	Fe^0	–
Fe^{SP} (ref. 70)	Laboratory	2 to 100	–	Crystalline	Fe^0	Magnetite, Maghemite
Pd/Fe^B (ref. 68) (bimetallic)	Laboratory	1 to 100	33.5	Amorphous	Fe^0	Goethite, Wustite

spectroscopy (EELS) and X-ray diffraction (XRD)[65,67–70] indicates that the RNIP and Fe^{SP} morphology was crystalline in structure compared with the round and amorphous shapes of Fe^B and Pd/Fe^B. Additionally, iron oxide phases such as magnetite and maghemite were observed as an outer shell of RNIP and Fe^{SP} (ref. 67,69,70), while no significant oxide shell was observed for Fe^B (ref. 65). Crystallinity and iron oxide shell critically impacts particle reactivity, the ability to utilise H_2, particle reactive life time, and therefore particle efficiency.[65,69] Control over the synthesis process and the resulting NP properties can therefore affect the cost and sustainability of groundwater treatment using NZVI.

The crystalline Fe^0 and an iron oxide shell (RNIP) reduced TCE *via* β-elimination with acetylene as the intermediate.[67] Further, at pH 7, a relatively small fraction of Fe^0 in RNIP was used to produce H_2, implying that RNIP has a good selectivity of electron utilisation for TCE dechlorination.[65,66] Consequently, RNIP has a relatively longer lifetime.[66]

The iron oxide shell grew as RNIP was oxidised (see Figure 8), however, making some fraction of Fe^0 in RNIP unavailable for dechlorination[67] and decreasing particle efficiency (only around 52% of the Fe^0 was used to degrade the contaminant). At pH 7, the measured surface area normalised pseudo-first order rate constant for TCE dechlorination was very similar to those reported for NZVI, implying that there is no true "nano effect" of RNIP reactivity, such as quantum effects when particles become nanosized.[71]

NZVI made from borohydride reduction did not have a significant iron oxide shell. However, at pH 7 a relatively large fraction of Fe^0 in Fe^B was used to produce H_2, implying that Fe^B has a poor selectivity of electron utilisation for dechlorination.[65] Fe^B has a much shorter reactive lifetime[67] in comparison to RNIP. But H_2 produced by Fe^B can be used for TCE dechlorination through a catalytic hydrodechlorination pathway. The ability of Fe^B to use H_2 for dechlorination is unique to these NPs and is attributed to the nanosized crystals (*ca.*1 nm) making up the particles.[65] This increases electron utilisation efficiency because electrons used to produce H_2 can ultimately be used by the particles to degrade TCE. The formation of iron oxide film around Fe^B was not observed and almost all of the Fe^0 in Fe^B was available for TCE dechlorination, *i.e.* high particle efficiency (*ca.* 92%).[65,67] One disadvantage of Fe^B particles, however, is their relatively short reactive lifetime in comparison to RNIP. Long reactive lifetimes are needed to make *in situ* treatment cost effective.[67] A reactive nanoparticle that can be delivered to the contamination source area has high selectivity and a long reactive lifetime is the ideal NP for *in situ* groundwater remediation. While not currently available, nanotechnology has the potential to provide such a particle in the future.

4.3   Bimetallic Particles for Transformations

Noble metals, such as Pd, Pt and Rh, are often doped to the surface of NZVI to enhance particle reactivity through catalytic pathways.[68,72,73] The structure of bimetallic NZVI particles is a reductive Fe^0 (or Zn) core with a surface layer of

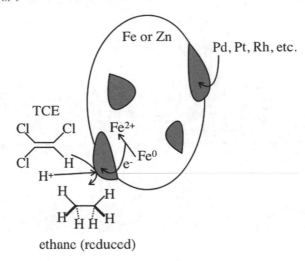

Figure 9 TCE reduction to ethane at the bimetallic NP catalyst site.

inert noble metals (see Figure 9). Because of the presence of noble metals, bimetallic particles can utilise H_2 for hydrodechlorination.[68] The surface area normalised pseudo-first order rate constant for TCE dechlorination using NZVI doped with Pd is ∼50 times greater than commercially available micron-sized Fe^0 particles.[68] Noble metals also protect the Fe^0 core from a non-productive oxidation by water or dissolved O_2 in water, thereby lengthening the reactive life time of the particles.[68] A significant problem with these types of bimetallic particles, however, is an observed decrease in reactivity over time[74] due to noble metal reactive surfaces being covered by a thick oxide layer. In addition, the health risk associated with metals such as Ni might make bimetallic NZVI a less environmental friendly choice compared with Fe^B or RNIP (ref. 67) particles.

4.4 Surface Modified Nanoparticles for in situ Groundwater Treatment

There are two general approaches for promoting the reaction between NZVI particles and contaminants in the subsurface (see Figure 10). Firstly, the nanoparticles can be delivered *in situ* to the TCE source area. Secondly, a reactive barrier can be placed in the ground to react with the dissolved contaminant plume as it passes out of the source zone (*e.g.* permeable reactive barrier or PRB).[75] The first approach requires NZVI to target the source (*e.g.* dense non-aqueous phase liquid or DNAPL) in the subsurface. Surface modification of NZVI is required to target DNAPL *in situ*. Three major modification alternatives include: (i) polymeric surface modification, (ii) emulsification and (iii) embedding or supporting NZVI to a carrier. Figure 10 shows the schematics of each approach.

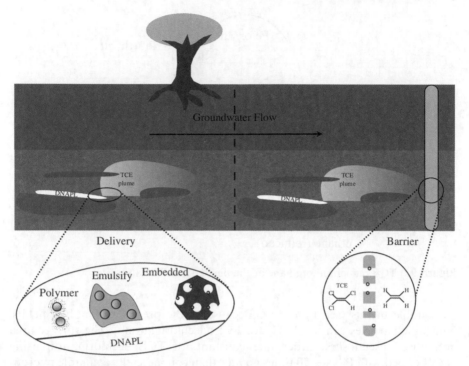

Figure 10 Application methods for *in situ* treatment of groundwater contamination source areas and plumes using reactive nanoparticles.

4.4.1 Polymeric Surface Modification

Nanoparticles without surface modification are effectively immobile in the subsurface and cannot be delivered to the contaminants. Polymers and polyelectrolytes enhance dispersion of nanoparticles *via* steric and electrosteric stabilisation, respectively, and improve mobility and contaminant targeting.[76,77] The efficacy of these polymers for these purposes depends on their chemical and structural complexity. Block copolymers allow for the addition of various functionalities to be built into the polymer blocks.[78–80] For example, one block can be designed to enhance the adsorption of polymer on nanoparticles, while another block can be designed to improve DNAPL targeting.[78] Foundational work[62,71] has allowed for polymeric surface modification of NZVI using block copolymers and homopolymers to become a common approach to controlling the mobility of these particles. This has resulted in the production of several types of commercially available polymer-modified NZVI.[81–87] Negatively charged polyelectrolytes or uncharged polymer are used for this purpose because aquifer materials at neutral pH are normally negatively charged.

Polymeric surface modification/functionalisation of NZVI can be achieved by (i) grafting polyelectrolytes from the surface of pre-synthesised nanoparticles, (ii) physisorption of polyelectrolytes onto the surface of pre-synthesised nanoparticles, or (iii) incorporating polyelectrolytes into the nanoparticles during the particle synthesis. The first approach uses advanced synthesis methods such as atom transfer radical polymerisation (ATRP)[80] to produce a dense polymer brush layer from initiators that are covalently bound to the nanoparticle surface. The dense polymer brush is theoretically predicted to provide superior enhancement of colloidal stability and nanoparticle transport. Adsorbing polymers onto nanoparticles (the second approach) is a less time- and material-intensive procedure than growing polymers from the particles and is desirable for applications requiring large quantities of reactive NPs, such as environmental remediation. Physisorption of homopolymers and block copolymers are reported to enhance the dispersion stability,[76] delivery[88] and NAPL emulsification.[89] A mixture of strong and weak polyelectrolytes, such as poly(styrenesulfonate) (PSS) and poly(acrylic acid)(PAA), together with bentonite-supported NZVI particles was also studied and reported to enhance NZVI transport, depending on mixture ratios.[83] The third approach, and one currently most popular in field implementation, is a single-step synthesis where a polymeric modifier interacts with the particle surface during nanoparticle synthesis.[82] Using carboxymethyl cellulose (CMC) as a stabiliser, Fe^{2+} ions are mixed with CMC to form Fe-CMC complexes prior to reduction using borohydride.[82] CMC-modified NZVI using this one pot synthesis is reported to enhance dispersion stability,[82] delivery and NAPL targeting.[84] Synthesising the NPs at the point and time of injection provides the freshest NPs with highest Fe^0 content to improve the efficiency of the materials for contaminant degradation.

4.4.2 Emulsified NZVI

Emulsification of NZVI is another modification of NZVI for active *in situ* groundwater remediation.[90] This approach delivers NZVI encapsulated in water-in-oil or oil-in-water emulsions directly into DNAPL.[91,92] For NZVI encapsulated in water-in-oil emulsion (see Figure 10), NZVI is in the interior of the emulsion, *i.e.* in the water phase encapsulated by the hydrophobic exterior of the oil phase, designed to deliver NPs into DNAPL. Surfactants such as Rhodopon (anionic; sodium laurel sulfate), Span 80 (nonionic; sorbitan monooleate) and Span 85 (nonionic; sorbitan trioleate) were used to make stable emulsion droplets.[92] On the other hand, for NZVI encapsulated in oil-in-water emulsion, NZVI is in the oil phase interior of the emulsion. Surfactants such as Aerosol MA (MA), Aerosol OT (AOT), oleic acid (OA) and Span 80 were used to both stabilise dispersions of NZVI in the oil phase and stabilise oil-in-water droplets. NZVI encapsulated in either water-in-oil and oil-in-water emulsions can improve NZVI delivery in the subsurface and may also provide DNAPL targeting; however, significant engineering hurdles remain to optimise this delivery technique.[91]

4.4.3 NZVI Embedded onto Carriers or Supports

Embedding NZVI on a porous carrier or support (*e.g.* silica or carbon black) is a modification approach designed to decrease NZVI aggregation and increase mobility in the subsurface. A carbothermal reduction process to synthesise carbon-supported NZVI (C-NZVI) has been designed.[93] This may be a low-cost synthesis alternative for NZVI by comparison with H_2 reduction or sodium borohydride reduction methods because chemical processing steps are not used (reducing high capital costs associated with producing large quantities of NZVI).[93] The iron reduction by carbon is an endothermic process yielding only gaseous byproducts, making it potentially scalable to large reactors and continuous processing. In this approach, Fe^{2+} or Fe^{3+} is adsorbed on or impregnated in carbon black. Fe-adsorbed carbon powder is subsequently heated to 600–800 °C under argon to obtain C-NZVI. The C-NZVI can also be functionalised by physisorbed polymer to further enhance its performance.[93]

Entrapping NZVI in porous silica nanoparticles was recently proposed[94,95] and may have two additional advantages for remediation purposes (aside from decreasing NZVI aggregation): the silica carrier size can be optimised for size, enabling improved porous media transport, and the silica carrier surfaces can be functionalised to promote contact with chlorinated organics.[95] Spherical (porous) silica particles are synthesised through an aerosol-assisted process[95] prior to the precipitation of NZVI in mesopores of the particles. Moreover, Pd or ethyl groups can be added to the surface of the NZVI to further enhance reactivity.[95]

4.4.4 Particles Embedded in Membranes

Embedding reactive particles as nanocomposites in membranes may provide several advantages over non-fixed particles. The surrounding membrane polymer limits oxidation of the Fe^0 particles because particles can be directly synthesised in the membrane.[96] Membrane material adsorbs contaminants to its surface, allowing for efficient delivery of contaminants to the particle surfaces.[96] Bimetallic particles with noble metals, such as Pd, are immobilised for possible reuse. Nanoparticles may be immobilised in polymers such as cellulose acetate,[97] polyvinylidene (PVDF)[98] and chitosan.[99] However, reaction rates may suffer from membrane incorporation.[96] While the use of reactive nanoparticles within barriers (*e.g.* PRBs) may show great potential so long as the selectivity of the NPs toward contaminants and the reactive lifetime of the materials are improved, their deployment *in situ* has not been explored to date.

5 Sustainability Challenges

Nanoscience-based water treatment processes are novel and often provide significant improvement over their conventional counterparts. Whether or not these processes are sustainable, however, remains an open question given the high energy cost associated with the synthesis of the materials. Nano-based processes must therefore be designed with a lifecycle analysis (LCA) in mind. LCA accounts for the acquisition of raw materials, manufacturing, use and end of life of the materials used within these processes.[100] The sum energy of the

LCA is the embodied energy of a material. Embodied energy must be weighted against any beneficial water treatment technology.

5.1 Raw Materials

Raw material acquisition can be energy intensive, depending on the source. Many nanomaterials based on metals and metal oxides originate from mining activities that are well known to be energy intensive and environmentally costly. For example, ilmenite, rutile or titanium slag must be mined in order to obtain the raw materials necessary for TiO_2 production.[101] Metal ores can be rare and expensive. For example, palladium only has an abundance of $6.3 \, ng \, g^{-1}$ in the earth's crust, so concentrating the materials to high purity for use carries with it a large thermodynamic penalty. Polymer materials that are used to embed or coat nanomaterials are often manufactured from hydrocarbons obtained from drilling and refining processes. Carbon black, a source of precursor carbon for fullerenes and CNTs,[102] is formed when carbons are incompletely combusted. On the other hand, some materials are renewable or recyclable. For example, cellulose acetate and chitosan polymers are derived from wood pulp and crustacean exoskeletons, respectively. Moreover, metals may also be obtained from recycling processes.

5.2 Manufacturing

Conversion of the raw materials into nanomaterials can be energy, environmentally and materials intensive. For example, TiO_2 uses an energy-intensive hydrolysis–calcination method, which uses a solvent of hydrochloric acid and requires high energy inputs.[103] Green chemistry practices call for reductions in emissions, toxicities and waste in manufacturing.[104] Data on nanomaterial manufacturing processes and their impacts are currently scarce. The proprietary nature of this information, as well as a lack of specific nanotech reporting requirements, pose formidable barriers to collecting the data from the companies carrying out the industrial processes in question. A body of research on nano-manufacturing processes, their resulting products and predicted production levels is beginning to emerge, however, which spans multiple disciplines including sociology, economics, engineering and the industrial R&D sector.[101,105–107]

One important consideration with respect to the impact of nanomaterial fabrication processes is, beyond their nano-scale products, the collateral impact of the production itself. Input products, by-products and waste streams will all play a role in the cumulative impact of industrial scale nano-production. One study focused on the insurance industry perspective, using an existing insurance company's risk algorithm to rank nanomaterial production processes in parallel with previously ranked conventional processes on a scale of 1–100 (100 being the most risky).[103] By virtue of being an expert-based relative ranking system, the XL Insurance system uses an expert chosen sub-set of key parameters describing chemical and processing characteristics to predict the relative

Table 4 Examples of relative risk scores for nanomaterial and conventional production processes (Adapted with permission from the American Chemical Society).[103]

Nano & Conventional Production Processes	Risk due to release from Accident/Catastrophic		Risk due to release during normal operations	
	Score	Culprit Material(s)	Score	Culprit Material(s)
Single-walled nanotubes	43	Carbon monoxide, Iron pentacarbonyl	23–34	Sodium hydroxide, Carbon monoxide
C_{60}	76	Benzene	40	Soot, Toluene
Q-dots	58	Hydrogen selenide	40	Hydrogen selenide, Carbon monoxide, Surfactant
Alumoxane	34	Acetic acid	29–40	Acetic acid, Aluminum oxide
Nano-TiO_2	62	Titanium tetrachloride	40–62	Titanium tetrachloride
Silicon wafers	53	Sulfuric acid	56	Arsine
Wine	39	Dithiocarbamate pesticides (zineb)	23	Sulfur dioxide
High density plastic (Polyolefin)	72	Titanium tetrachloride, Vinyl acetate	62	Titanium tetrachloride
Automotive lead–acid batteries	58	Lead dioxide	51	Sulfuric acid, Lead monoxide
Refined petroleum	76	Benzene, Toluene	67	Benzene, Toluene, Xylenes
Aspirin	58	Phenol, Toluene	23	Phenol

risk. These parameters are chosen on the basis of what most affects the risk posed by the overall process: carcinogenicity, persistence, explosive tendency, atomic weight, pressure levels, temperature levels, emission rates during production, *etc*. The highest ranking or most "risky" material involved with each of the processes strongly dictates the resulting score for the process as a whole. Table 4, adapted from this study, shows the relative ranks of five nanofabrication processes and six conventional processes based on the results from the insurance algorithm.

5.3 Use and End of Life

Nanomaterial use and fate after the useful life of the materials is an important and expanding topic in environmental engineering. Applications of nanomaterials, along with their use in consumer products, are growing rapidly.[108] In the context of water treatment applications, it is important to consider that nanomaterial incorporation may lead to release during the life cycle of those products and will increase the occurrence of manufactured nanomaterials in the environment. For example, the leaching of silver NPs used as antimicrobial agents in cloth was recently reported.[109] Unintentionally or intentionally applied NPs might come into contact with biological receptors.[110] Therefore, it is important to design processes while mindful of the ultimate fate of the nanomaterials contained within.

6 Conclusions

Our planet faces serious challenges stemming from population growth, health, water supply, energy and pollution issues. Scientists and engineers are faced with finding solutions to these challenges that are sustainable. A significant area where many of these challenges overlap is water treatment technologies. Nanoscience could lead to creation of treatment processes that improve upon existing technologies in both performance and sustainability by decreasing raw material needs, while increasing reaction rates and surface chemistry specificity. However, any potential improvements need to be tempered by accounting for the entire life cycle cost of a technology through LCA. LCA involves accounting for raw material acquisition, manufacturing, use and disposal of the materials involved in the processes. The risks and potential benefits of nanomaterial utilisation in water treatment therefore garners attention from the scientific community, governmental agencies and public stakeholders.

Acknowledgements

This material is based upon work supported by the National Science Foundation (NSF) and the Environmental Protection Agency (EPA) under NSF Cooperative Agreement EF-0830093, Center for the Environmental Implications of NanoTechnology (CEINT). Any opinions, findings, conclusions or recommendations expressed in this material are those of the author(s) and do not necessarily reflect the views of the NSF or the EPA. This work has not been subjected to EPA review and no official endorsement should be inferred.

References

1. WHO, *The World Health Report 1995 – Bridging the Gaps,* WHO, Geneva, 1999.
2. ASCE, 2010, http://www.asce.org/reportcard/2009/grades.cfm, accessed Februrary 24, 2010.
3. M. Roco, *J. Nanopart. Res.*, 2004, **6**, 1–10.
4. M. Auffan, J. Rose, J.-Y. Bottero, G. V. Lowry, J.-P. Jolivet and M. R. Wiesner, *Nat. Nanotechnol.*, 2009, **1–8**.
5. G. Boorman, V. Dellarco, J. Dunnick, R. Chapin, S. Hunter, F. Hauchman, H. Gardner, M. Cox and R. Sills, *Environ. Health Perspect.*, 1999, **107**, 207–217.

6. S. W. Krasner, H. S. Weinberg, S. D. Richardson, S. J. Pastor, R. Chinn, M. J. Sclimenti, G. D. Onstad and A. D. Thruston, *Environ. Sci. Technol.*, 2006, **40**, 7175–7185.
7. G. Nangmenyi and J. Economy, in *Nanotechnology Applications for Clean Water*, ed. N. Savage, M. Diallo, J. Duncan, A. Street and R. C. Sustich, William Andrew Inc., Norwich New York, 2009, 3–15.
8. J. Morones, J. Elechiguerra, A. Camacho, K. Holt, J. Kouri, J. Ramirez and M. Yacaman, *Nanotechnology*, 2005, **16**, 2346–2353.
9. O. Choi and Z. Hu, *Environ. Sci. Technol.*, 2008, **42**, 4583–4588.
10. F. Zeng, C. Hou, S. Wu, X. Liu, Z. Tong and S. Yu, *Nanotechnology*, 2007, **18**, 055605.
11. V. K. Sharma, R. A. Yngard and Y. Lin, *Adv. Colloid Interface Sci.*, 2009, **145**, 83–96.
12. P. Jain and T. Pradeep, *Biotechnol. Bioeng.*, 2005, **90**, 59–63.
13. H. Le Pape, F. Solano-Serena, P. Contini, C. Devillers, A. Maftah and P. Leprat, *Carbon*, 2002, **40**, 2947–2954.
14. G. Nangmenyi, W. Xao, S. Mehrabi, E. Mintz and J. Economy, *J. Water Health*, 2009, **7**, 657–663.
15. V. A. Oyanedel-Craver and J. A. Smith, *Environ. Sci. Technol.*, 2008, **42**, 927–933.
16. A. R. Bielefeldt, K. Kowalski and R. S. Summers, *Water Res.*, 2010, **43**, 3559–3565.
17. M. R. Hoffmann, S. T. Martin, W. Y. Choi and D. W. Bahnemann, *Chem. Rev.*, 1995, **95**, 69–96.
18. Q. Li, W. Pinggui and J. K. Shang, in *Nanotechnology Applications for Clean Water*, ed. N. Savage, M. Diallo, J. Duncan, A. Street and R. C. Sustich, William Andrew Inc., Norwich, New York, 2009, 17–37.
19. E. Stathatos, H. Choi and D. D. Dionysiou, *Environ. Eng. Sci.*, 2007, **24**, 13–20.
20. H. Choi, E. Stathatos and D. Dionysiou, *Thin Solid Films*, 2006, **510**, 107–114.
21. R. Asahi, T. Morikawa, T. Ohwaki, K. Aoki and Y. Taga, *Science*, 2001, **293**, 269–271.
22. C. Burda, Y. Lou, X. Chen, A. Samia, J. Stout and J. Gole, *Nano Lett.*, 2003, **3**, 1049–1051.
23. J. Li and J. Z. Zhang, *Coord. Chem. Rev.*, 2009, **253**, 3015–3041.
24. Q. Li, W. Liang and J. K. Shang, *Appl. Phys. Lett.*, 2007, **90**, 063109.
25. J. Chen, D. Ollis, W. Rulkens and H. Bruning, *Water Res.*, 1999, **33**, 661–668.
26. P. Wu, R. Xie, J. A. Imlay and J. K. Shang, *Appl. Catal., B*, 2009, **88**, 576–581.
27. P. Wu, R. Xie and J. K. Shang, *J. Am. Ceram. Soc.*, 2008, **91**, 2957–2962.
28. E. Hotze, J. Labille, P. J. J. Alvarez and M. Wiesner, *Environ. Sci. Technol.*, 2008.
29. K. Pickering and M. Wiesner, *Environ. Sci. Technol.*, 2005, **39**, 1359–1365.

30. J. Lee, J. Fortner, J. Hughes and J. Kim, *Environ. Sci. Technol*, 2007, **41**, 2529–2535.
31. A. Badireddy, E. Hotze, S. Chellam, P. J. J. Alvarez and M. Wiesner, *Environ. Sci. Technol.*, 2007, **41**, 6627–6632.
32. E. M. Hotze, A. R. Badireddy, S. Chellam and M. R. Wiesner, *Environ. Sci. Technol.*, 2009, **43**, 6639–6645.
33. J. Lee, Y. Mackeyev, M. Cho, D. Li, J.-H. Kim, L. J. Wilson and P. J. J. Alvarez, *Environ. Sci. Technol.*, 2009, **43**, 6604–6610.
34. I. Escobar, E. Hoek, C. Gabelich, F. DiGiano, Y. Le Gouellec, P. Berube, K. Howe, J. Allen, K. Atasi, M. Benjamin, P. Brandhuber, J. Brant, Y. Chang, M. Chapman, A. Childress, W. Conlon, T. Cooke, I. Crossley, G. Crozes, P. Huck, S. Kommineni, J. Jacangelo, A. Karimi, J. Kim, D. Lawler, Q. Li, L. Schideman, S. Sethi, J. Tobiason, T. Tseng, S. Veerapanemi and A. Zander, *J. Am. Water Works Assn.*, 2005, **97**, 79–89.
35. M. Clark, S. Allgeier, G. Amy, S. Chellam, F. DiGiano, M. Elimelech, S. Freeman, J. Jacangelo, K. Jones, J. Laine, J. Lozier, B. Marinas, R. Riley, J. Taylor, M. Thompson, J. Vickers, M. Wiesner and A. Zander, *J. Am. Water Works Assn.*, 1998, **90**, 91–105.
36. S. Komarneni, *J. Mater. Chem.*, 1992, **2**, 1219–1230.
37. B.-H. Jeong, E. M. V. Hoek, Y. Yan, A. Subramani, X. Huang, G. Hurwitz, A. K. Ghosh and A. Jawor, *J. Membr. Sci.*, 2007, **294**, 1–7.
38. M. L. Lind, A. K. Ghosh, A. Jawor, X. Huang, W. Hou, Y. Yang and E. M. V. Hoek, *Langmuir*, 2009, **25**, 10139–10145.
39. H. S. Lee, S. J. Im, J. H. Kim, H. J. Kim, J. P. Kim and B. R. Min, *Desalination*, 2008, **219**, 48–56.
40. S. Y. Lee, H. J. Kim, R. Patel, S. J. Im, J. H. Kim and B. R. Min, *Polym. Adv. Technol.*, 2007, **18**, 562–568.
41. J. S. Taurozzi, H. Arul, V. Z. Bosak, A. F. Burban, T. C. Voice, M. L. Bruening and V. V. Tarabara, *J. Membr. Sci.*, 2008, **325**, 58–68.
42. V. Thomas, M. M. Yallapu, B. Sreedhar and S. K. Bajpai, *J. Biomater. Sci. Polym. Ed.*, 2009, **20**, 2129–2144.
43. B. Hinds, N. Chopra, T. Rantell, R. Andrews, V. Gavalas and L. Bachas, *Science*, 2004, **303**, 62–65.
44. N. Ma, X. Fan, X. Quan and Y. Zhang, *J. Membr. Sci.*, 2009, **336**, 109–117.
45. H. Choi, E. Stathatos and D. Dionysiou, *Appl. Catal., B*, 2006, **63**, 60–67.
46. X. Zhang, T. Zhang, J. Ng and D. D. Sun, *Adv. Funct. Mater.*, 2009, **19**, 3731–3736.
47. P. Fortina, L. Kricka, S. Surrey and P. Grodzinski, *Trends Biotechnol.*, 2005, **23**, 168–173.
48. M. Majumder, N. Chopra, R. Andrews and B. Hinds, *Nature*, 2005, **438**, 44–44.
49. J. Holt, H. Park, Y. Wang, M. Stadermann, A. Artyukhin, C. Grigoropoulos, A. Noy and O. Bakajin, *Science*, 2006, **312**, 1034–1037.
50. M. Yu, H. H. Funke, J. L. Falconer and R. D. Noble, *Nano Lett.*, 2009, **9**, 225–229.

51. M. S. Mauter and M. Elimelech, *Environ. Sci. Technol.*, 2008, **42**, 5843–5859.
52. T. A. Hilder, D. Gordon and S. -H. Chung, *Small*, 2009, **5**, 2183–2190.
53. C. Y. Won and N. R. Aluru, *J. Am. Chem. Soc.*, 2007, **129**, 2748–2749.
54. H. Guo, Y. Wyart, J. Perot, F. Nauleau and P. Moulin, *J. Membr. Sci.*, 2010.
55. T. V. Cech, *Principles of Water Resources: History, Development, Management, and Policy,* John Wiley & Sons Inc, Hoboken, NJ, 2005.
56. C. Hogue, *Chem. Eng. News*, 2005, **83**, 14.
57. H. Stroo, M. Unger, C. Ward, M. Kavanaugh, C. Vogel, A. Leeson, J. Marqusee and B. Smith, *Environ. Sci. Technol.*, 2003, **37**, 224A–230A.
58. EPA, *List of Contaminants & their MCLs*, 2010; http://www.epa.gov/safewater/contaminants/index.html, Accessed 22/02/2010.
59. M. R. Higgins and T. M. Olson, *Environ. Sci. Technol.*, 2009, **43**, 9432–9438.
60. D. Song, M. Conrad, K. Sorenson and L. Alvarez-Cohen, *Environ. Sci. Technol.*, 2002, **36**, 2262–2268.
61. Y. Lin, X. Cui and J. Bontha, *Environ. Sci. Technol.*, 2006, **40**, 4004–4009.
62. T. Phenrat and G. V. Lowry, in *Nanotechnology Applications for Clean Water*, ed. N. Savage, D. Mamadou, J. S. Duncan, A. Street and R. C. Sustich, William Andrew, Norwich NY, 2009, 249–267.
63. Y. Lin, D. Choi, J. Wang and J. Bontha, in *Nanotechnology Applications for Clean Water*, ed. N. Savage, M. Diallo, J. Duncan, A. Street and R. C. Sustich, William Andrew Inc., Norwich, New York, 2009, 179–189.
64. B. Gu, W. Dong, G. Brown and D. Cole, *Environ. Sci. Technol.*, 2003, **37**, 2291–2295.
65. Y. Liu, H. Choi, D. Dionysiou and G. V. Lowry, *Chem. Mater.*, 2005, **17**, 5315–5322.
66. Y. Liu and G. V. Lowry, *Environ. Sci. Technol.*, 2006, **40**, 6085–6090.
67. Y. Liu, S. A. Majetich, R. D. Tilton, D. S. Sholl and G. V. Lowry, *Environ. Sci. Technol.*, 2005, **39**, 1338–1345.
68. W. Zhang, *J. Nanoparticle Res.*, 2003, **5**, 323–332.
69. J. T. Nurmi, P. G. Tratnyek, V. Sarathy, D. R. Baer, J. E. Amonette, K. Pecher, C. Wang, J. C. Linehan, D. W. Matson, R. L. Penn and M. D. Driessen, *Environ. Sci. Technol.*, 2005, **39**, 1221–1230.
70. D. R. Baer, P. G. Tratnyek, Y. Qiang, J. E. Amonette, J. Linehan, V. Sarathy, J. T. Nurmi, C.-M. Wang and J. Antony, in *Environmental Applications of Nanomaterials*, ed. G. E. Fryxell and G. Cao, Imperial College Press, London, 2007.
71. G. V. Lowry, in *Environmental Nanotechnology: Applications and Impacts of Nanomaterials*, ed. M. R. Wiesner and J. -Y. Bottero, McGraw-Hill, New York, 2007.
72. W.-X. Zhang, C.-B. Wang and H.-L. Lien, *Catal. Today*, 1998, **40**, 387–395.
73. B. Schrick, J. L. Blough, A. D. Jones and T. E. Mallouk, *Chem. Mater.*, 2002, **14**, 5140–5147.

74. M. M. Scherer, S. Richter, R. L. Valentine and P. J. J. Alvarez, *Crit. Rev. Environ. Sci. Technol.*, 2000, **30**, 363–411.
75. A. D. Henderson and A. H. Demond, *Environ. Eng. Sci.*, 2007, **24**, 401–423.
76. T. Phenrat, N. Saleh, K. Sirk, H.-J. Kim, R. D. Tilton and G. V. Lowry, *J. Nanoparticle Res.*, 2008, **10**, 795–814.
77. G. J. Fleer, M. A. Cohen Stuart, J. M. H. M. Scheutjens, T. Cosgrove and B. Vincent, *Polymers at Interfaces,* Chapman & Hall, New York, USA, 1998.
78. N. Saleh, T. Phenrat, K. Sirk, B. Dufour, J. Ok, T. Sarbu, K. Matyjaszewski, R. D. Tilton and G. V. Lowry, *Nano Lett.*, 2005, **5**, 2489–2494.
79. G. J. Fleer, M. A. Cohen Stuart, J. M. H. M. Scheutjens, T. Cosgrove and B. Vincent, *Polymers at Interfaces,* Chapman & Hall, London, USA, 1993.
80. K. Matyjaszewski and J. Xia, *Chem. Rev.*, 2001, **101**, 2921–2990.
81. K. W. Henn and D. W. Waddill, *Remediation*, 2006, **16**, 57–77.
82. F. He, D. Zhao, J. Liu and C. B. Roberts, *Ind. Eng. Chem. Res.*, 2007, **46**, 29–34.
83. B. W. Hydutsky, E. J. Mack, B. B. Beckerman, J. M. Skluzacek and T. E. Mallouk, *Environ. Sci. Technol.*, 2007, **41**, 6418–6424.
84. F. He, M. Zhang, T. Qian and D. Zhao, *J. Colloid Interf. Sci.*, 2009, **334**, 96–102.
85. R. L. Johnson, G. O. Johnson, J. T. Nurmi and P. G. Tratnyek, *Environ. Sci. Technol.*, 2009, **43**, 5455–5460.
86. A. Tiraferri and R. Sethi, *J. Nanoparticle Res.*, 2009, **11**, 635–645.
87. N. Saleh, K. Sirk, Y. Liu, T. Phenrat, B. Dufour, K. Matyjaszewski, R. D. Tilton and G. V. Lowry, *Environ. Eng. Sci.*, 2007, **24**, 45–57.
88. N. Saleh, H.-J. Kim, T. Phenrat, K. Matyjaszewski, R. D. Tilton and G. V. Lowry, *Environ. Sci. Technol.*, 2008, **42**, 3349–3355.
89. N. Saleh, T. Sarbu, K. Sirk, G. V. Lowry, K. Matyjaszewski and R. Tilton, *D. Langmuir*, 2005, **21**, 9873–9878.
90. C. A. Ramsburg, K. D. Pennell, T. C. G. Kibbey and K. F. Hayes, *Environ. Sci. Technol.*, 2003, **37**, 4246–4253.
91. N. D. Berge and C. A. Ramsburg, *Environ. Sci. Technol.*, 2009, **43**, 5060–5066.
92. C. Geiger, C. A. Clausen, K. Brooks, C. Clausen, C. Huntley, L. Filipek, D. R. Reinhart, J. Quinn, T. Krug, S. O'Hara and D. Major, in *Chlorinated Solvent and DNAPL Remediation: Innovative Strategies for Subsurface Cleanup*, ed. S. M. Henry and S. D. Warner, American Chemical Society, Washington DC USA, 2002, 132–140.
93. L. B. Hoch, E. J. Mack, B. W. Hydutsky, J. M. Hershman, J. M. Skluzacek and T. E. Mallouk, *Environ. Sci. Technol.*, 2008, **42**, 2600–2605.
94. J. Zhan, T. Zheng, G. Piringer, C. Day, G. L. McPherson, Y. Lu, K. Papadopoulos and V. T. John, *Environ. Sci. Technol.*, 2008, **42**, 8871–8876.

95. T. Zheng, J. Zhan, J. He, C. Day, Y. Lu, G. L. McPherson, G. Piringer and V. T. John, *Environ. Sci. Technol.*, 2008, **42**, 4494–4499.
96. S. M. C. Ritchie, in *Nanotechnology Applications for Clean Water*, ed. N. Savage, M. Diallo, J. Duncan, A. Street and R. C. Sustich, William Andrew Inc., Norwich, New York, 2009, 293–309.
97. L. Wu and S. M. C. Ritchie, *Environ. Prog.*, 2008, **27**, 218–224.
98. G. K. Parshetti and R.-A. Doong, *Water Res.*, 2009, **43**, 3086–3094.
99. S.-J. Wu, T.-H. Liou and F.-L. Mi, *Bioresour. Technol.*, 2009, **100**, 4348–4353.
100. R. Lankey and P. Anastas, *Ind. Eng. Chem. Res.*, 2002, **41**, 4498–4502.
101. C. Robichaud, A. Uyar, M. Darby, L. Zucker and M. Wiesner, *Environ. Sci. Technol.*, 2009, **43**, 4227–4233.
102. Y. Ju-Nam and J. R. Lead, *Sci. Total Environ.*, 2008, **400**, 396–414.
103. C. Robichaud, D. Tanzil, U. Weilenmann and M. Wiesner, *Environ. Sci. Technol.*, 2005, **39**, 8985–8994.
104. M. J. Eckelman, J. B. Zimmerman and P. T. Anastas, *J. Ind. Ecol.*, 2008, **12**, 316–328.
105. L. G. Zucker and M. R. Darby, *Nanoscience and Nanotechnology: Opportunuties and Challenges in California*, California Council on Science and Technology, Sacramento, CA, 2004.
106. E. Fauss, University of Virginia, 2008.
107. *US Pat.*, 6 440 383, 2002.
108. Project on Emerging Nanotechnologies, *Nanotechnology Consumer Products Inventory*, 2010; http://www.nanotechproject.org/, Accessed 24/02/2010.
109. T. M. Benn and P. Westerhoff, *Environ. Sci. Technol.*, 2008, **42**, 4133–4139.
110. M. R. Wiesner, G. V. Lowry, K. L. Jones, M. F. Hochella, R. T. Di Giulio, E. Casman and E. S. Bernhardt, *Environ. Sci. Technol.*, 2009, **43**, 6458–6462.

Subject Index

Abstraction, 80, 88
  licence trading (water trading), 88
ACQWA project, 17
ADAM project, 18
Adaptation to Climate Change, 14
Adapting to climate change, 10
Advanced oxidation processes, 142
Africa, 25–27, 31–32, 34–40, 44
African water resources, 25
Agricultural water efficiency, 85
Agri-food industry, 89
Analytical methods, 50, 57
Animal toxicological studies, 119
Anthropogenic pressure, 2–3, 5, 8
Appropriate quality water, 87
AQUAMONEY project, 18
AquaStress, 17, 19
Aquifer Storage, 37
Australian
  Drinking Water Guidelines,
    133–134
  Guidelines for Water Recycling,
    122, 132

Benchmarking, 87
Bimetallic particles, 152–153, 156
Biodiversity, 4
Blue water, 80, 82–83
Boron nitride, 147

Carcinogenic risk, 121
Catchment Abstraction Management
  Strategy (CAMS), 88
Chemical
  contaminants, 116
  Monitoring Activity (CMA), 51
  risk assessment, 118

CIRCE project, 17
Climate
  and hydrogeological maps of
    Africa, 36
  change, 1–4, 10–11, 15, 25–26,
    28
    Adaptation in the Water Sector,
      18
    adaptation policies, 10
    impacts, 4
    in 'hot spots', 11
    in Africa, 31
    mitigation, 3
    scenarios, 15
  modelling uncertainties, 34, 45
  projections, 30
  Water project, 19
CORFU project, 20
Corporate social justice, 109

Dangerous Substances Directive,
  54
Deprivation, 102, 104, 107
  level, 107
Desertification, 4
Di(2-ethylhexyl)phthalate (DEHP),
  75
Disinfection, 138, 140
Drinking water supplies, 27
Droughts, 19, 27
  in Africa, 27

Ecological footprint, 79
Electrically switched ion exchange,
  149
End of life, 158
ENSEMBLES project, 15

Environmental
  justice, 94
  Quality Standards, 52, 55, 64
  water quality standards, 9
EU
  Environmental Quality Standards,
    50
  policies, 5
  QA/QC Directive, 58
  water policy, 9
EURO-LIMPACS project, 16
European
  Centre for the Validation of
    Alternative Methods, 125
  Drought Policy, 19
  Environment Agency, 14
  Water Framework Directive
    (WFD), 50–51
Extreme events, 32

FLASH project, 20
Flood
  Action Programme, 10–11
  defence
    funding, 96
    resources, 97
  events, 36
  impacts, 96
  research, 19
  risk management, 20
Flooding, 96
Floods Directive, 10
FLOODsite project, 19
Food security, 43, 89
Fossil fuel emissions, 32
Freshwater resource management, 5
Fullerene photochemistry, 144
Fullerenes, 143, 145
Functionalised fullerenes, 144

Global
  carbon cycle, 30
  Circulation Models, 29
  mean temperature rise, 33
Glyphosate, 71, 73
Green water, 80

Groundwater
  contaminants, 148
  contamination, 154
  development, 26
  pollution, 86
  recharge, 37, 45
  remediation, 148–149, 155
  resources, 38, 44
  treatment, 153

HighNoon project, 17
Hollow fibre membrane, 148
Hurricane Katrina, 96
Hydrological cycle, 1–2, 14, 17
Hyogo Framework for Action, 13

IMPRINTS project, 20
Improved
  rural water supplies, 25, 27, 44
    in Africa, 37, 44
Inactivation of pathogens, 141
Income deprivation, 103, 105
  by deciles, 105
Index of multiple deprivation (IMD),
  102
Indicator
  chemicals, 126
  compounds, 127
Integrated water resources
  management (IWRM), 12, 21
Interface between science and policy,
  22
Intergovernmental Panel on Climate
  Change (IPCC), 2
International Water Supply and
  Sanitation Decade, 95
IPCC
  Fourth Assessment
    of Climate Change, 28
    Report, 32
  scenario storylines, 28
  Special Report on Emissions
    Scenarios, 33
  statements about climate change
    and water, 6
  Technical Paper on Water, 4

Irrigated
  agricultural production, 89
  agriculture, 80
  food production, 78
  vegetables, 81
Irrigation, 26, 43–45, 78, 80–86
  abstractions, 78
  application, 86
  efficiency, 85
  licences, 85
  practices, 86

Joint Monitoring Programme, 26

Life-cycle analysis, 138, 156
Limit of Quantification (LOQ), 64
Long-lived greenhouse gases, 29

Managing abstraction, 87, 88
Mecoprop, 71–72
Membrane, 145–147, 156
Metered properties, 104
MIRAGE project, 19
Monitoring
  frequencies, 56–57
  programme, 54–55
Multiple deprivation, 103
Municipal wastewater, 115

Nanocomposite membranes, 145
Nanocomposites, 146, 156
Nanomaterial fabrication, 157
Nanomaterials, 138–139, 157
Nanoparticle photocatalysts, 143
Nanotube
  embedded membranes, 147
  pores, 147
National Water Council, 98
NeWater project, 18

Ofwat (the Water Services Regulation
  Authority), 99
Oligodynamic effect, 140
Organochlorine pesticides, 59
Organotin compounds, 75
Oxidation technologies, 138

Pentabromodiphenylether, 62
Perchlorate removal, 150
Perchlorates, 149
Photocatalytic semiconductors, 142
Planned potable water recycling,
  115
Policy responses, 4
Polluter Pays Principle, 99
Polycyclic Aromatic Hydrocarbons
  (PAHs), 60, 73
Polyelectrolytes, 155
Polymer brush, 155
Population growth, 26, 28, 43
Potable water recycling, 114,
  116, 118
Potomac Estuary Experimental Water
  Treatment Plant, 125
Priority
  action substances (PAS), 64–65
  substances, 52, 59, 72
Probability density functions, 128
PRUDENCE project, 15

Quality of Water Resources, 38

Rainfall
  events, 32
  patterns, 3, 30, 32
    in Africa, 32
Raw materials, 157
Reactive oxygen species, 142–143
Red book, 118
Regional
  Climate
    Model, 15
    Scenarios, 15
  Water Authorities, 98
Relative risk, 122, 159
Remediation technologies, 138
Reverse osmosis, 129–130
Risk
  assessment, 9, 118, 122, 132, 135
  matrix, 134
River basin management, 4, 10
  planning, 12
Rural water supply in Africa, 34

Safe drinking water concentration,
    120
Salinisation of groundwater, 2
Sampling strategy, 55
Sanitation, 38, 95, 139
    in rural Africa, 38
Science – policy links, 21
Semiconductor photocatalysis, 142
Short–Chain Chlorinated Paraffins,
    62
Silica nanoparticles, 157
Silver
    nanocomposites, 146
    nanoparticles, 140–141, 146
        inactivation of bacteria, 141
Singlet oxygen, 143
Social justice, 93–94, 98–99, 109
    ramifications, 99
Socio-economic development, 29
Storage reservoirs, 87
Substances of priority concern, 51
Summer
    precipitation, 110
    temperatures, 110
Surface
    modification, 154
    water
        monitoring, 64
        resources in Africa, 35
Sustainable abstraction, 88

Tampa Water Resource Recovery
    Project, 126
Technology choices for rural Africa,
    40
Titanium dioxide, 142
Toxicity, 119, 123–124, 126, 134–135
    equivalents, 134
    testing, 124, 126
        techniques, 135
Toxicological
    assessments, 120
    databases, 120
    evaluation, 119
    testing, 123
Tributyltin, 60, 74

UK water companies, 97
UN Economic Commission for
    Europe, 12
Uncertainties in
    African climate projections, 30
    Climate Projections, 29
Unimproved
    sources, 27
    water
        sources, 25, 44
        supplies, 38
United
    Kingdom Water Industry
        Research, 99
    Nations
        Framework Convention on
            Climate Change, 10
        International Strategy for
            Disaster Reduction, 13
Unmetered properties, 104
US
    National Toxicological Program
        Interagency Centre for the
        Evaluation of Alternative
        Toxicological Methods, 125
    Safe Drinking Water Act, 118

Virtual embedded water, 79

WATCH project, 16
Water
    and Sewage Companies, 97
    card, 108
    Charges Equalisation bill, 98
    -climate interactions, 14
    debt, 99–102, 107
    efficiency, 83
    footprint, 78–79, 83, 86, 88, 90
    Framework Directive, 2, 5, 21,
        50–51, 53–54, 111
    industry debts, 101
    justice, 93
    management, 3
    Only Companies, 97
    security, 26, 34, 39, 45
        in Africa, 34, 39

Western Corridor Recycled Water
   Project, 116
Wet season, 36, 38
WFD
   Common Implementation
      Strategy, 51
   objectives, 10

River Basin Management Plans, 21

XEROCHORE
   project, 19
   Support Action, 19

Zeolites, 145–146